The Accelerating Universe

Infinite Expansion, the Cosmological Constant, and the Beauty of the Cosmos

Mario Livio

John Wiley & Sons, Inc.
New York • Chichester • Weinheim • Brisbane • Singapore • Toronto

To Sharon, Oren, and Maya, who I hope will find beauty in their own lives

Published by John Wiley & Sons, Inc.
Published simultaneously in Canada

Library of Congress Cataloging-in-Publication Data

Livio, Mario
The accelerating universe : infinite expansion, the cosmological constant, and the beauty of the cosmos / Mario Livio.
p. cm.
Includes index.
ISBN 0-471-32969-X (cloth: alk. paper)
ISBN 0-471-39976-0 (paper)
1. Cosmology. I. Title.
QB981.L57 1999
523.1—dc21 99-22278

Printed in the United States of America

10 9 8 7 6 5 4 3 2 1

Contents

Foreword

When the history of ideas is written four hundred years from now, the twentieth century will be known as the dawn of the great scientific idea of the origin of a changing universe that is still evolving. It recycles hydrogen and helium through the stars in its galaxies and manufactures therefrom the heavier chemical elements that somehow can organize themselves into complex structures that contemplate themselves. Self-contemplation is but one of many miraculous things that the natural chemical elements do over the age of the universe.

These ideas of origins and evolution have become part of scientific literacy of our age. The major ideas of this twentieth-century synthesis center around the understanding of the laws of physics, the life history of the stars, and the consequences of that history for mankind in an expanding universe.

This book is about how such grand ideas have come about using the cosmological discoveries made in astronomy and physics. But it is also about much more. It concerns the philosophy of science, of how one judges if a theory is fundamental and likely to last, about truth, about the definition of beauty, and about the connection of science to art, literature, music, and the human endeavor.

Imagine two authors, one an art and literature fanatic, and the other a theoretical physicist who is heavily involved in cosmology. Imagine a decision to write a book jointly that would attempt to bridge the two cultures. Could the two such authors make a synthesis that would ring true, each to his or her own tradition, yet also true to the other's quite different creative style? Now imagine that the two authors are not two, but are the same person. What you are about to read is that synthesis. Its author is that art fanatic and theoretical physicist. What could such a book be like?

Near the middle third of the just-past century, a series of highly influential, semipopular, yet scientifically astute books were published that had a profound influence on the very young (and not so young), many of whom prepared themselves for a scientific life because of them. Such books, especially by Sir James Jeans and Sir Arthur Eddington, were important for astronomy. Perhaps the most popular was *The Mysterious Universe* (1930) by Jeans, which Tallulah Bankhead, the American actress of multiple reputations, said contains "what every girl should know." Other books of the same sort were *The Stars in Their Courses* (1931), *Through Space and Time* (1933), and the favorite of this writer, *Physics and Philosophy* (1943), all by Jeans. Similar central books by Eddington were *The Nature of the Physical World* (1931), *Science and the Unseen World* (1929), *The Expanding Universe* (1933), and again a favorite of mine, *Space, Time, and Gravitation* (1920), which opened up general relativity and Riemannian manifolds to the layman. More recently, the books by Fred Hoyle such as *The Nature of the Universe* (1950), *Frontiers of Astronomy* (1955), and *Highlights of Astronomy* (1975) had the same merit.

This book by Mario Livio is of that type. Not only is it a book on the new astronomy and cosmology, but it also is a book on the "old" philosophy—of art, and of culture. Livio's deep purpose is to explore the meaning of beauty in science, using the developments in twentieth-century cosmology as the subject with which to discuss a concept of beauty in general and beauty in science in particular. The author's premise is that "beauty" is an essential ingredient in all truly successful theories in science, and especially "true" theories of the nature of the universe. The aim of the book is to discuss whether the laws of physics are actually determined by aesthetic principles.

Livio believes they are, and makes a compelling case throughout the book. He first defines what he means by beauty. He then shows that so much of the modern standard model of cosmology conforms to his stated definition. In the process he describes almost every aspect of both the old and the new cosmology. This is the heart of the book, both for the young would-be scientist and for the fascinated layman. In the process he shows what beauty means in art, describing many paintings by many artists to illustrate both the beauty and the philosophy of human artistic endeavor.

Said only this way, most hard-bench, experimental reductionist scientists would likely be put off by such an attempt. We scientists have generally been trained to regard art, literature, poetry, and anything else to do with the "humanities" as subjective and therefore not amenable to assessment by the scientific method. On the other hand, science, and especially physics and astronomy, is widely thought to refer to nature independently of subjective thought, feeling, and categories made by the mind. Beauty often is also placed outside that regime. This book is a revelation on how, and at what price (if any), this notion of objectivity with no concern for beauty is incompatible with modern science, especially in cosmology.

In chapter 2, the author defines beauty as he will use it throughout the book. Although he agrees with the often-quoted definition by many that "beauty symbolizes a degree of perfection with respect to some ideal," yet this definition is too diffuse and oversimplified to be of use. He also rejects the misquoted (out of context) definition by Keats that "beauty is truth, truth beauty," and states that beauty defined by the poets is often a dangerous thing, as Paris found by his association with Helen.

Livio does insist, as do many of the greatest scientists throughout history, that the laws of physics are exquisitely beautiful. Before defining beauty, he convinces us that in physics, the explanations of what once were mysteries but which now are "understood" is often ineffably more beautiful than the questions, and that "beauty in physics and cosmology is not an oxymoron." To proceed in a way that is precise enough so that in later chapters he can combine the concept of beauty with the most esoteric and profound elements of modern cosmology, Livio sets out his definition of beauty in this early, introductory chapter.

Three requirements must be fulfilled if a theory is to be judged beautiful:

1. It must describe a symmetry (or a series of symmetries); otherwise its predictions will not be invariant to the two simplest transformations as to place (i.e., space) or time (it makes no difference if I do an experiment now or next week), or at a much deeper level by circumstances in the equations (i.e., such as coordinate transformations).
2. It must have simplicity in the sense of reductionism (i.e., many questions can be replaced by very few more basic questions that can be solved as puzzles and not be left as mysteries).
3. A theory must obey a generalized Copernican principle, by which is meant that we, or the circumstances, are nothing special, in time, space, or category.

This last is probably the deepest requirement, and has the most profound implication for the ultimate cosmological mystery, described in the penultimate chapter as the Nancy Kerrigan question, whose solution can be given in terms of "quintessence," which in fact provides an adherence to the postulated cosmological aesthetic principle. This, and the final chapter, provide the climax to the full story of cosmology that has been set out in chapters 3 through 8.

These six chapters contain the description of much of cosmology of the twentieth century. The discovery of the expansion in 1929 is in chapter 3, with its implications for the hot big bang origin and the resulting prediction and subsequent discovery of the relic radiation and the formation of the earliest elements out of the gluon glue (once called the yelm), predictions made already in 1945 by George Gamow, Ralph Alpher, and Robert Herman. Here, the more modern ideas that combine the high-energy particle physics of the 1980s with its Higgs fields and the grand unification theories of the four forces are made easy by the many surprising analogies from art, literature, and underground cultural jokes combined with the straightforward style of the author. The book itself has charm as well as beauty.

Chapter 4 combines the problem of the missing mass with the question of the fate of the universe via the omega parameter. The method of inferences from indirect data is illustrated with the examples of studies of the authenticity of Rembrandt paintings and the deductions by Sherlock Holmes on the alcoholism and marital neglect of a client, and the discovery of the omega minus particle (no relation to the omega density parameter) independently by Yuval Ne'eman after its prediction by Murray Gell-Mann. All this is tied up with beauty in the simplicity of the deductions as they ultimately relate to the fate of the expansion.

The richness of the narrative and the fun of the author is kept throughout the book as it lays out even the most complicated of the cosmological problems. The prime example is in chapters 5 and 6 where non-Euclidean geometry, curved space as the manifestation of gravity, and inflation via the behavior of the inflaton fields following the Planck time are all explained better than in any other semipopular book on this most profound of the cosmological subjects. These ideas are at the heart of the question of origins and the "cause" of the expansion, as well as the explanation of the dominance of matter over antimatter.

Chapters 5, 6, and 7 are titled "Flat Is Beautiful," "When Inflation Is Good," and "Creation." These contain the essence of modern quantum cosmology as a theory of the origin, evolution, and fate of the universe (or many universes in the variant called "eternal inflation," as invented by Alex Vilenkin and Andrei Linde). Here the author is most persuasive and powerful in attempting to fulfill his promise to convince us that "never in the history of physics have aesthetic arguments played a more dominant role than in the attempts to answer the question of whether the universe will expand forever."

The central premise of the book is that good or true theories must, and ultimately always are, beautiful in their entirety. Here, Livio makes the distinction between "beautiful" and "elegant." An incorrect theory can be elegant in its argumentation and yet not be beautiful if it is not "simple," or does not deny special places or special circumstances (the Copernican egalitarianism). It is claimed that if parts of such elegant theories remain ugly in the anti-Copernican sense, then the great scientist will go after the ugliness (any way you wish to define ugliness) and will generally find a beautiful solution that is more true, if a solution in fact exists. Many scientists have said the same, often differently, but with the same meaning. Livio gives many examples.

The final two chapters are the best to cement the argument that the Copernican principle must be obeyed for a beautiful theory. Livio asks whether the question of life, and/or its uniqueness (or otherwise) in the universe, has a bearing on the Copernican principle. What if we could prove that life is unique to us here? We then would be something special and one of a kind. Would that destroy cosmology as "beautiful" because it would endow our existence with a special status? It is this writer's opinion that the two final chapters are not only the most profound in the book, but also that Livio's solution to the dilemma of uniqueness and the apparent contradiction of his requirement for the Copernican criterion is astoundingly beautiful in itself.

However, the reader must get there alone. I will not reveal Livio's clever solution except to mention as a teaser the several ideas of a nonconstant cosmological constant (the quintessence field); the anthropic "principle", Brandon Carter's argument that life must be unique to Earth because of the near equality of two large numbers; Livio's refutation of the argument; the changing views of what constitutes beauty, illustrated by what various societies regard as the ideal ratio of hip to waist girth in the human anatomy; and the falling of the Nancy Kerrigan problem to the advantages of introducing quantum quintessence, making nonsense of the question "why now?" In this way, Livio argues that cosmological beauty can be restored.

Those who read this book will have been on as authoritative a journey as can be taken through modern cosmology as is possible for interested nonexperts. Those travelers will have seen the nuts and bolts of the 1930 side of Hubble's early discoveries in cosmology. They will have witnessed the marriage of that early field of the observational study of the expanding universe with nuclear astrophysics by the discovery of the relic radiation in the 1960s and its connection with the hot early universe. The discovery of quasars, pulsars, gravitational lenses, and radio, X-ray, and gamma ray galaxies sent the field still further into high-energy astrophysics and ultimately, beginning in the 1980s, into the lap of the ultra-high-energy particle physicists and the deeper lap of theoreticians of the gauge principle and of quantum field theory and quantum gravity. In short, the whole of modern cosmology has been revealed and exposed in this book.

Furthermore, the reader will have met such artists as van Gogh, Dali, Mondrian, Edvard Munch, Ferdinand Hodler, Vermeer, Kadinsky, van der Weyden, Cézanne, and five times more, and such writers as Dante, Shakespeare, Pope, Lizzie Siddal, Shikibu, Antoine de Saint-Exupéry, William Cowper, Timothy Leary, and many more, and at least fifty current scientific cosmologists and their work. What more could you ask? (The jokes are also splendid.) This is a book with charm, beauty, elegance, and importance.

Allan Sandage
April 1999

Preface

I am sure that even if I came up with a dozen good definitions of "beauty," the reader could still challenge most of them. The reaction would probably be even more skeptical if I announced that I intend to discuss beauty in physics and cosmology. Yet the original meaning of the word *aesthetic* was "to perceive by the senses." This does not distinguish between our reaction to a work of art or to the wonders of the universe. Therefore, this book deals with my two great passions—for science, in particular astronomy and astrophysics, and for art. As such, the book does not attempt to represent a comprehensive description of all aspects of modern astrophysics; rather, it follows a central theme of beauty in theories of the universe.

I hope that science lovers and art lovers alike will find this book both exciting and fully accessible.

It would be absolutely impossible for me even to attempt to thank all the many colleagues who have, directly and indirectly, contributed to the contents of this book. I would like, however, to extend special thanks to Allan Sandage, Alex Vilenkin, Lee Smolin, and Mark Clampin for their careful reading of an early version of the manuscript and for their valuable comments. I am also grateful to Nathan Seiberg and Ronen Plesser for their helpful comments. An ancient Hebrew saying states, "I have learned from all my teachers, but most of all from my students." There is no doubt that the contents of this book have largely been influenced and shaped by the many popular lectures I have given in recent years and by the feedback I received from the audiences.

I would like to thank my agent, Susan Rabiner, for her encouragement and advice, especially in the early stages of writing. Sarah Stevens-Rayburn, Barbara Snead, and Elizabeth Fraser provided me with indispensable library and linguistic assistance (always with a smile). I am grateful to Sharon Toolan, Jim Ealley, Ron Meyers, and Dorothy Whitman for their technical support during the preparation of the manuscript. This book owes much of its present form to my editor at John Wiley & Sons, Emily Loose. Her careful critical reading, insistence on clarifications, and imaginative suggestions have been absolutely invaluable.

Many years ago I saw a book in which the author wrote: "My wife had nothing to do with it!" While it is true that my wife, Sofie Livio, had nothing to do with the actual writing, this book would *never* have been written without her infinite patience and continuous support during the writing process. For this I am forever grateful.

1

Prologue

How did our universe begin? and how will it end? Progress toward answering these questions has usually been incremental, relying on continuous improvements in observational techniques, coupled with advances in theoretical understanding.

Every few decades, however, an observation reveals something so unexpected that it forces theoreticians back to the drawing board. Even more rarely, the findings of such an observation are so startling that they qualify as a revolution. In 1998, cosmology—the study of the universe at large—experienced such a revolution.

Two teams of astronomers presented strong evidence suggesting that the expansion of our universe is accelerating. Thus, not only may our universe expand forever, it may do so at ever-increasing speeds. As if this were not enough, these new findings also suggest that the behavior of the universe is dominated neither by the energy of ordinary matter nor even by the energy of more exotic matter (the existence of which is suggested by theories of the basic forces in the universe). Rather, the 1998 observations suggest that our universe is dominated by the energy of empty space.

When in the sixteenth century Copernicus (and later Galileo) dethroned the earth from its central position, this was (and still is) considered a revolution. In Bertolt Brecht's play *Life of Galileo,* an aged cardinal says: "I am informed that Signor Galilei transfers mankind from the center of the universe to somewhere on the outskirts. . . . Is it conceivable that God would trust this most precious fruit of his labor to a minor frolicking star?" The new observations imply that far from being central, the earth's role (and that of the stuff the earth is made of) in the workings of the universe is vanishingly small compared to the role of the vacuum.

Apart from their obvious significance to cosmology, the newly emerging data have implications at a deeper, more philosophical level. As I will show in this book, they pose a serious challenge to a centuries-old assumption that has proved extremely effective, that the ultimate theory of the universe must be *beautiful.* This is the kind and magnitude of challenge sometimes referred to as a paradigm shift.

Let me start with an examination of the science that is responsible for this new revolution—astronomy. It is not always easy to identify what it is that makes certain disciplines more attractive or "interesting" than others. Certainly among the sciences, astronomy has always enjoyed a quite privileged status, in igniting the imagination of even nonprofessionals. The romantic appeal of the stars, coupled with the natural curiosity about the origin of the universe, and of life within it, have combined to make astronomy almost irresistible.

The fascination with astronomy has been further enhanced by the fact that modern observations, in particular with the *Hubble Space Telescope,* have produced a gallery of images that are a true feast for the eyes. In fact, some images of nebulae—glowing clouds of gas and dust—and of violently colliding galaxies generate in viewers an emotional response similar to that experienced when contemplating the most extraordinary works of art. I would like to carry the analogy between viewing the universe and viewing works of art further in this book, and explain the place of beauty, and of aesthetic appreciation, in our understanding of the universe. One may ask, for example, how viewers of, say, a particular painting, can enhance their aesthetic appreciation of the work. They might do this by: (1) viewing other paintings by the same artist, in order to be able to place the painting in a broader context; (2) learning more about the artist's personality and life, in order to become better acquainted with possible drives/motivations; and (3) investigating the background of the particular painting, to generate the type of empathy that is often required for a truly deep emotional reaction.

A perfect example for this process is provided by the common reaction to a poster of a painting that has been hanging on the wall in my office for the past seven years. After the initial "Wow! What a beautiful painting," scientists who walk into my office invariably start to ask me numerous questions about it. The painting is *Ophelia,* first presented in 1852 by the Pre-Raphaelite painter John Everett Millais (the painting itself is in the Tate Gallery in London).

The Pre-Raphaelite Brotherhood was the name adopted by a small group of English painters in 1848. They were linked by a revolutionary spirit and a strong reaction to what they regarded as meaningless academic subject painting. The name of the brotherhood implies a challenge to the then supreme authority of the High Renaissance painter Raphael and a return to medieval art, in a quest for ideal beauty.

The painting shows Ophelia floating down a flower-strewn stream, apparently oblivious, in her madness, of her imminent death by drowning (a wonderful description of the painting and its history can be found in *The Pre-Raphaelites,* edited by Leslie Paris). Death, in particular of youth, in relation to grieving broken hearts was a recurring motif in Victorian culture in general and in paintings of the Pre-Raphaelites in particular. It is hard to imagine a more appropriate representation of this tragic motif than through the character of Ophelia, who, driven insane by the murder of her father by her lover Hamlet, drowns herself in the stream. In Shakespeare's words (act 4, scene 7 of *Hamlet*):

There, on the pendent boughs her coronet weeds
Clambering to hang, an envious sliver broke;
When down her weedy trophies and herself
Fell in the weeping brook. Her clothes spread wide,
And, mermaid-like, awhile they bore her up;
Which time she chanted snatches of old tunes,
As one incapable of her own distress,
Or like a creature native and indued
Unto that element; but long it could not be
Till that her garments, heavy with their drink,
Pull'd the poor wretch from her melodious lay
To muddy death.

The painting contains numerous flowers and plants that are all charged with symbolism, emblematic of Ophelia's fate. For example, floating pansies, associated with thought and vain love; a necklace of violets around Ophelia's neck, symbolizing faithful chastity and the death of the young; daisies, symbolizing innocence; and forget-me-nots, simply carrying the meaning of their name.

Millais painted the riverbank first, on site, on the Hogsmill River near Ewell in Surrey. He started working on the background on July 2, 1851, and finished in mid October, making it highly unlikely for all the flowers in the painting to have been in bloom simultaneously. Millais himself wrote on the process: "I am threatened with a notice to appear before a magistrate for trespassing in a field and destroying the hay; likewise by the admission of a bull in the same field after the said hay be cut; am also in danger of being blown by the wind into the water, and becoming intimate with the feelings of Ophelia when that lady sank to muddy death."

The model for Ophelia was Elizabeth (Lizzie) Siddal, a lover and later wife of the Pre-Raphaelite leader Dante Gabriel Rossetti. Millais himself bought the ancient dress, described in his words as "all flowered over in silver embroidery," in which Lizzie posed. The figure was painted during the winter months, with Lizzie inside a tin bathtub filled with water, which was kept warm by lamps underneath. Millais was painting with such concentration that at one time the lamps went out unnoticed and Lizzie caught a severe cold.

Sadly, Lizzie Siddal's own fate bears a tragic resemblance to that of Ophelia. Around 1860, already seriously ill, she wrote a poem called "A Year and a Day," a part of which reads:

> The river ever running down
> Between its grassy bed
> The voices of a thousand birds
> That clang above my head
> Shall bring to me a sadder dream
> When this sad dream is dead.

She died in February 1862 of a self-administered overdose of the tincture of opium called laudanum.

I have gone to some length in describing what, I am sure, is obvious to many: that uncovering several strata of background of a work of art can lead to an entirely different level of aesthetic perception and appreciation.

By analogy, let me now describe briefly the process that astronomers and astrophysicists followed, once confronted with the first images of these spectacular nebulae, in particular of the type known as planetary nebulae. Incidentally, an image of such a nebula hangs on my wall, side by side with *Ophelia*.

As a first step, astronomers collected more data, to determine whether the nebulae are a rare, unusual phenomenon in the universe, or whether they are rather common. This data collection process resulted in the discovery of about two thousand planetary nebulae. Further study revealed that embedded within each of these diffuse nebulae there is essentially always a hot central star. These findings immediately implied that we are dealing here with some relatively ordinary outcome of *stellar evolution*—the process defining the life of a star, from birth to death. Furthermore, by observing individual central stars, astronomers determined that they are always of relatively low mass, similar to that of our sun. A combination of observations of the properties of the nebulae and the stars, and theoretical models simulating the evolution of stars, then established the fact that these planetary nebulae represent the end stages in the lives of stars similar to the sun. We are thus witnessing here the future fate of our own sun. In their old age, the stars, which are expanded at that point to giant dimensions, shed off their loosely bound outer layers, exposing a hot central core. That core irradiates the ejected gases in a final blaze of glory and causes them to fluoresce like bright, colorful neon lights.

In spite of this general understanding, a few surprises were still to come. When high-resolution, extremely sharp images of the nebulae became available, astronomers discovered that the nebulae do not look at all, as everybody expected, like round spherical shells. Rather, like snowflakes, each one is complex and different. Only after several more years astronomers noticed that the different shapes can be grouped into some general classes, such as "round," "elliptical," "butterfly," and so on. Eventually, this led to *one* underlying basic model, which can, at some level, explain the formation of *all* the observed shapes of nebulae. In very simple terms, the proposed model works as follows.

Most stars are not single, like our sun, but rather come in pairs. As the more massive star of the pair evolves, it expands to become a giant, with dimensions hundreds of times larger than the sun. This giant can engulf the companion star and literally swallow it into its outer layers—its envelope. As the companion revolves inside the giant's envelope it acts like an eggbeater, causing those layers to rotate rapidly. Consequently, the giant star ejects slowly its outer layers in the form of a doughnut (a torus) around its equator. Once the envelope is lost, the hot core of the giant is exposed. This core ejects more dilute but fast material in all directions (spherically). However, the doughnut acts like a corset, not allowing the fast material to expand at the equatorial waist, thus forcing it to blow two bubbles in the directions of the poles, like those you would obtain by squeezing in the middle of an elongated balloon.

Depending on how dense and how thick the doughnut is, this dynamical model can produce a variety of shapes, ranging from slightly oval to an hourglass. Our understanding of the physical processes that shape planetary nebulae has thus finally reached a stage where it can be described by the verses of the eighteenth-century poet Alexander Pope:

> Where order in variety we see,
> And where, though all things differ, all agree.

As we shall see later in this book, this general idea of identifying one law that can explain a variety of phenomena is one of the most fundamental goals of modern physics.

The above two examples, of the Millais painting and of the planetary nebulae, demonstrate that there do exist some similarities between the pursuit of a deeper aesthetic meaning in a work of art and the search for the underlying "truth" in a physical phenomenon. Ultimately, one searches for one fundamental idea that forms the basis for an explanation. However, there are also many important differences between art and science, and one of the main ones, from the point of view of the present book, is the following. As I will explain in chapter 2, the behavior and evolution of the universe are governed by a set of laws, which we call the *laws of physics.* These laws are truly universal, in the sense that the same laws apply to the entire observable universe. Therefore, if intelligent extraterrestrial creatures really exist (a topic discussed in chapters 8 and 9), they should deduce the same laws from their part of the universe as we do from ours. This means that such creatures would be able, once the language barrier is overcome, to actually *understand* our science. The same, however, cannot be concluded about our art, because art in general is *not* derived from a set of universal laws (and it's a good thing, too!).

The last sentence is not meant to be understood as if aesthetic principles that are applied to works of art *never* rely on some general, more universal rules. For example, there have been attempts to set up a canon for a perfectly pleasing proportion in a work of art. The best known of these is the *Golden Section.* This is a name given in the nineteenth century to a proportion derived from a division of a line (see Figure 1a), in which the ratio of the larger segment (*a* in the figure) to the smaller segment *(b)* is the same as that of the sum of the segments (*a* plus *b*) to the larger segment *(a)*. In the plane, the Golden Section is given by the line that connects two identical Golden Section points on opposite sides of the rectangle shown in Figure 1b (this is also known as a Golden Rectangle). This particular line division has apparently been known at least since the time of ancient Egypt. It was discussed at length by the mathematician Luca Pacioli, who was a close friend of Leonardo da Vinci and Piero della Francesca, in his 1509 book *De Divina proportione* (for which Leonardo da Vinci prepared sixty drawings). As a number, the ratio that represents the Golden Section is given approximately by 1.618 . . . ; it is exactly equal to one-half of the sum of 1 and the square root of 5. It has been argued

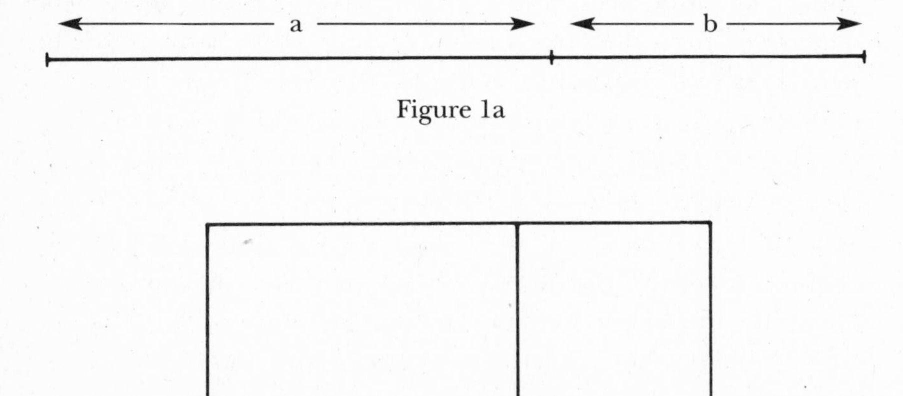

Figure 1a

Figure 1b

that the use of the Golden Section proportion in works of art is aesthetically superior to all other proportions, because it offers an asymmetric division, or an element of tension, which at the same time maintains a certain balance. There are numerous examples of the use of the Golden Section in art, starting from the Parthenon in ancient Greece, through paintings by Botticelli to modern art. For example, in his famous painting *The Sacrament of the Last Supper* (which is in the National Gallery of Art in Washington, D.C.), Salvador Dali made use of the Golden Section both by making the entire painting a Golden Rectangle and by using other internal Golden Rectangles for the positioning of the various figures. Other examples of the use of Golden Sections for positioning include *The Card Players,* by Paul Cézanne, and *Spatial Concept—Expectations,* one of the famous slit canvases by the Italian (born in Argentina) abstract modern artist Lucio Fontana.

It is interesting that outside of art, some of the mathematical properties of the Golden Section are associated with natural phenomena. The mathematician Leonardo of Pisa, called Fibonacci, discovered in the twelfth century a series of numbers 1, 1, 2, 3, 5, 8, 13, 21, 34 . . . (in which, starting with the third number, each number is the sum of the two preceding numbers), which is intimately related to the Golden Section. The ratio of each number in the series to the one preceding it rapidly approaches the Golden Section. For example, 5/3 is equal to 1.666 . . . ; 8/5 is equal to 1.60; 13/8 is equal to 1.625; 21/13 is equal to 1.615, thus getting closer and closer to 1.618. . . . As was noted above, many phenomena in nature exhibit Fibonacci patterns. These include the positions of leaves along a stalk (the phenomenon known as phyllotaxis) and the arrangements of some flower petals. Also, the points that mark the Golden Sections on successively embedded rectangles (Figure 2) all lie on a spiral pattern (known as a logarithmic spiral), which is extremely common in nature, from the chambered spiral shells of tropical mollusks to the observed shapes of some giant galaxies.

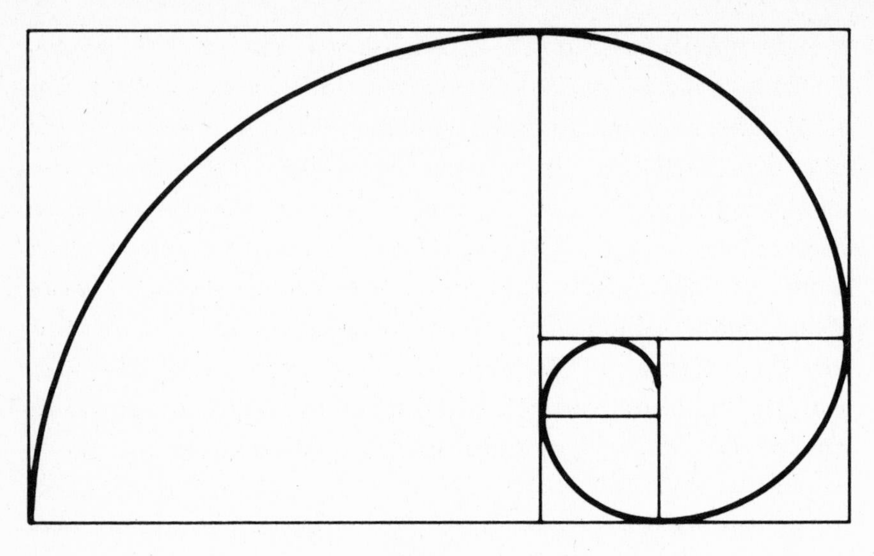

Figure 2

Music appreciation also contains some "universal" rules. I am not referring here to compositional structures, but rather to a physical basis that underlies all musical forms. One can analyze the *spectrum* of music—the distribution of the intensity of the sounds over the different frequencies or pitches. Such an analysis was performed in 1978 by two Berkeley physicists, Richard Voss and John Clarke. They discovered that a wide variety of musical styles is characterized by the same spectral shape. The type of spectrum they found is known as "one over f noise." It is distinguished by the fact that there is no special time interval over which sounds are correlated; rather, correlations exist over all intervals. This results in a pleasing sound sequence that is neither too boring nor too random. On a much less scientific basis, simply having had my three children grow up in the house, my impression is that as we grow older our taste evolves from appreciating the less predictable music toward appreciating the more predictable one.

The mere existence of such mathematically based concepts as the Golden Section and one over f noise serves as testimony to the fact that beauty in art can also contain some truly universal elements. However, one has to fully acknowledge the fact that the history of art fortunately does contain, and in all likelihood will continue to contain, many examples of extraordinary works that did not and will not conform to any type of universal canon.

While beauty in art is not governed by a set of laws, I will argue that the laws of physics *are* actually largely determined by aesthetic principles. At the heart of this book is what I hope will be a fascinating look at a concept dear to most physicists: the idea that one test of whether a physical theory of the universe is correct is its beauty. One of my goals is to identify exactly what those aesthetic standards are and are not, and to explain why they and not others are the important ones. I will then show how these standards helped physicists settle on a theory that attempts to explain the origin and evolution of the universe, from its smallest to its largest structures. In particular, I will concentrate on one of the basic questions intriguing astronomers and cosmologists today: Will our presently expanding universe continue to expand forever? Or will the expansion come to a halt and the universe start contracting?

As will be explained in detail in chapter 5, cosmologists have long known that the answer to this question depends (mainly) on the density of matter (or energy) in the universe. If the universe is denser than some critical value, the gravitational attraction of the matter within the universe is sufficient to brake the current expansion and allow collapse to begin.

If, on the other hand, the density of matter in the universe is lower than the critical value, then gravity is too weak to stop the expansion, and it will continue forever, with a speed that while possibly decreasing slowly, will never approach zero (the speed could also be increasing).

In the borderline case in which the density of matter in the universe is *exactly equal* to the critical density, the universe will expand forever, but with an ever-decreasing sluggish speed, reaching zero at some infinitely distant time in the future.

Astronomers have been trying for years to determine which of the above scenarios truly represents our universe, by attempting to measure the deceleration of the cosmic expansion (the denser the universe, the faster it brakes). These measurements rely on distant beacons of light. For example, because of their enormous power, supernovae—extremely energetic explosions marking the end of stellar lives—prove to be very useful in this regard, since they can be seen huge distances from our own Milky Way galaxy. If the expansion is indeed slowing down, objects at great distances—that is, objects that emitted their light long ago—should appear to be receding faster than the normal expansion law predicts. Furthermore, the greater the deceleration (i.e., the higher the density), the larger this difference should be. Hence, attempting to measure the deviations of very distant supernovae from normal expansion can help astronomers determine the density of the universe.

But long before cosmologists had conclusive data on which of the above three alternatives would prove correct, many (if not most) cosmologists maintained that the most likely cosmological model is in fact the one in which the density of the universe is *exactly equal* to the critical density. This prejudice was not formed on the basis of compelling theoretical models, but rather largely on the basis of *aesthetic* arguments.

Therefore, this book is about *beauty*. But it does not deal with the usual questions that are associated with the creation of art, its meaning, and its influence. Rather, the first part of the book explores beauty as an essential ingredient in theories of the nature of the universe. However, the second part of this book, one based on the very recent astronomical data, sheds new light on everything described in the first part. Why? Because cutting-edge findings—for example, from those distant supernovae—suggest that the density of *matter* in the universe, a key factor in determining the ultimate fate of the universe, may be smaller than critical after all. Furthermore, these new observations indicate that the universe may be experiencing the action of some form of cosmic repulsion that causes the expansion to accelerate.

This new information immediately suggests that the universe may expand forever, *at an ever-increasing rate,* and this will in itself influence future cosmological research in innumerable ways.

But the possibility that the density of matter is smaller than critical and the potential existence of this cosmic repulsion also contradict existing notions of what a "beautiful" theory of the universe should look like.

One may claim that in the visual arts, in some sense, the viewer actually *contributes* to the creation of a work of art, and thus to its beauty. If this were not the case, then given the fact that the optics of vision is the same in all viewers, all of the artistic experiences would have been identical. It is through psychological responses that involve creativity that differences among the viewers' perceptions are generated.

As we shall see in this book, a question arises of whether humans play a similar role in "determining" the properties of the universe, and thus its beauty. Namely, could it be, for example, that some properties of our observable universe are what they are because these are prerequisites for *our* existence? Could it be that of all the possible universes, one in which humans are most likely to emerge on some planet has to have a density that is less than critical? As I will argue, this possibility, if true, violates some of the principles required for beauty. Hence, we are facing the frightening challenge of having possibly to question, or maybe even abandon, the idea of a beautiful theory of the universe. The latter, if confirmed, may prove as traumatic to physicists today as abandoning the idea of our special place in the universe was to Galileo's inquisitors.

Are there ways out of this conundrum? Does a yet more global and beautiful theory exist? Or could our ideas of beauty itself change?

This book will attempt to examine these and other questions, each of which will result in a tour of intriguing topics in modern astrophysics and cosmology.

2

Beauty and the Beast

What is beauty? What is it that makes certain works of art, pieces of music, landscapes, or the face of a person so appealing to us that they give us an enormous sense of excitement and pleasure? This question, with which many philosophers, writers, psychologists, artists, and biologists have struggled at least since the time of Plato, and which led (among other things) to the birth of the notion of aesthetics in the eighteenth century, is still largely unanswered. To some extent, all of the classical approaches to beauty can be summarized by the following (clearly oversimplified) statement: Beauty symbolizes a degree of perfection with respect to some ideal. It is strange, though, that something which has such an abstract definition can cause such intense reactions. For example, some accounts claim that the Russian writer Dostoyevsky sometimes fainted in the presence of a particularly beautiful woman.

In spite of some changes in taste over the centuries (and some obvious differences among different cultures), the perception of what is beautiful is very deeply rooted in us. It suffices to look at a few paintings like Botticelli's *Primavera* or Leighton's *Flaming June*, or at a majestic mountain landscape, to realize this.

The term *beautiful* has evolved from being identified with "good" and "real" or "truthful" in ancient Greece to a quality that is confined merely to its effects on our senses, in the second half of the eighteenth century. It is interesting to note, though, that the approach of the philosophical book of Proverbs in the Bible has been rather dismissive and closer to the more modern definition in its assertion: "Charm is deceitful, and beauty is vain."

But, even if only affecting our senses, the effects of beauty should not be underestimated. The ancient Greeks certainly did not underestimate them. Greek mythology contains the famous story of the goddess Eris, who, insulted because she was not invited to the wedding party of King Peleus and the sea nymph Thetis, decided to take revenge by throwing into the banquet hall a golden apple that carried this inscription: "For the Fairest." After a long debate among the goddesses, the choice was narrowed down to three contestants for the title: the goddesses Hera, Athena, and Aphrodite. The matter was brought for a decision to Zeus who (very wisely) passed the task on to Paris, the son of the King of Troy. As it turned out, Paris's job was reduced to an evaluation of the bribes offered to him by each of the goddesses.

Hera whispered to him that she would make him the Lord of Europe and Asia; Athena promised him victory over the Greeks; and Aphrodite made him an offer he could not refuse—she promised him that the most beautiful woman in the entire world would be his. Paris gave the apple to Aphrodite, an action that can only be described as a mistake of historic proportions. The most beautiful woman on earth was Menelaus's wife, Helen, whose face "could launch a thousand ships." The end of the story is tragic. After Paris kidnapped Helen and brought her to Troy, a fierce and bitter war broke out, which led eventually to the total destruction of Troy.

Helen's beauty is described as being so intense, and its effects so devastating, that when Menelaus decides to execute Helen, a Trojan hero's mother forces him to swear that he will perform the execution without looking into Helen's eyes, because "through the eyes of men she controls them and destroys them in the same way that she destroys cities."

Some speculate that Helen's beauty was of the cold, unapproachable type, and that its overwhelming effects stemmed from the fact that Helen appeared as unattainable as the understanding of the concept of beauty itself.

Sometimes one can understand a certain concept or quality better by examining something that epitomizes the opposite. This notion is partly responsible for such pairings as heaven and hell, Dr. Jekyll and Mr. Hyde, Stan Laurel and Oliver Hardy, and, indeed, as the title of this chapter implies, Beauty and the Beast. So far I have only discussed beauty, but where is the "beast"?

The beast, in this case, is—physics! To many of my personal friends and to a large number of students of humanities whom I have met over the years, there is nothing more remote from the notion of beauty and more antithetical, from the point of view of the sensation that it induces in them, than physics. In fact, the disgust and fear that physics stimulates in some people is rivaled only by their feelings toward cockroaches. In an article in the *Sunday Times* (quoted in a BBC lecture by Richard Dawkins), columnist A. A. Gill compared observations in the sky to movie and theater stars by saying, "There are stars and there are stars, darling. Some are dull, repetitive squiggles on paper, and some are fabulous, witty, thought-provoking . . ." Believe it or not, those "dull, repetitive squiggles" represented the discovery of *pulsars,* objects so dense that one cubic inch of their matter weighs a billion tons, and that take a fraction of a second to rotate, instead of the earth's twenty-four hours!

I hope that this book will convince even skeptics that "beauty in physics and cosmology" is not an oxymoron. I remember a certain scene in the movie *Good Morning, Vietnam* in which a soldier is asked to which unit he belongs. His answer, "military intelligence," provokes an immediate reaction from the general: "There is no such thing!" In relation to science, the English poet Keats virtually accused Newton of ruining the beauty of the rainbow by his theoretical explanations of how it is formed, using the laws of optics. In Keats's words:

Philosophy will clip an Angel's wings
Conquer all mysteries by rule and line,
Empty the haunted air, and gnomed mine—
Uneave a rainbow . . .

Incidentally, some readers may find the latter story surprising, given that Keats is often quoted as having said: "Beauty is truth, truth beauty." In fact, Keats said no such thing. It is what he said the Grecian Urn depicts, in his criticism of works of art that deliberately eliminate existing difficulties of life.

Keats's complaint merely reflects the general feeling that magicians' tricks often lose their charm once we know how they are performed. However, in physics, very often the explanation is even more beautiful than the question, and even more frequently, the solution to one puzzle helps uncover an even deeper and more intriguing mystery. I therefore hope to be able to demonstrate that reactions like Keats's merely represent a misunderstanding that is based on false myths.

What Is Beautiful?

Definitions are always difficult, especially when we are dealing with something that is (largely) subjective. In this sense, even the definition in the *Oxford English Dictionary*—"Impressing with charm the intellectual or moral sense, through inherent fitness or grace"—which surely does not involve all the intricacies originating from philosophical interpretations, is not particularly useful.

I will attempt to make my task easier by answering at this point a much simpler question. Since my goal is to discuss beauty in physics and cosmology, I will address the question: When does *a physicist* feel that a physical theory is beautiful?

Any endeavor aimed at answering this question is bound to result in the use of an entire vocabulary of concepts, mostly borrowed from the arts. The list of such concepts may include symmetry, coherence, unity, harmony, and so on.

Probably not all physicists agree on which subset of concepts from this list should be used. However, I will argue later that at least three requirements are absolutely essential and *must* be fulfilled:

1. Symmetry
2. Simplicity
3. The Copernican principle

A fourth element, elegance, is also considered by some to be an important ingredient of a beautiful theory, but, as I will explain later, I do not consider it essential.

I realize that at this stage the meanings of all of these concepts are vague at best (if not totally obscure), but I will now explain in some detail what I mean by each one of them.

1. Symmetry: When Things That Might Have Changed Do Not

Everyone is familiar at least with symmetries of pictures, objects, or shapes. For example, our face and body have an almost exact bilateral symmetry. What this means is that if we reflect each half of our face, we obtain something that is nearly identical to the original (strangely enough, this is true even for the one-eyed giant Cyclops whom Ulysses encountered in his travels).

Some shapes are symmetric with respect to rotation. For example, a circle drawn on a page remains the same if we rotate the page on the desk.

Other arrangements are symmetric under certain displacements or translations. For example, if we stood in front of some of the row houses in Baltimore facing one unit, and someone were to displace the entire row by one unit, we would not notice any difference. Similarly, if we look at one Campbell soup can in a painting containing rows of identical cans by the pop artist Andy Warhol, and the painting is shifted slightly sideways or upward, we see an identical picture.

Notice that in all of these examples the object or shape *did not change* when we performed the symmetry operation, reflection, rotation, or translation.

The association of symmetry with beauty does not require elaborate explanations. Anyone who ever looked through a kaleidoscope has experienced the sensation of beauty that symmetry inspires. In fact, the word *kaleidoscope* itself comes from the Greek words *kalos,* which means "beautiful," and *eidos,* which means "form" (*skopeein* means "to look").

It is important to emphasize, though, that when the concept of symmetry is introduced into physics, it is not the symmetry of shapes that we are interested in but rather *the symmetry of the physical laws.* As we shall soon see, in this case, too, the symmetry is associated with things that do not change. In order to explain this concept better, let me first describe briefly the nature of these entities that we call laws of physics.

The laws of physics, sometimes referred to as the laws of nature, represent attempts to give a mathematical formulation to the behavior that we observe all natural phenomena to obey. For example, in classical physics, Newton's universal law of gravitation states that every particle of matter in the universe attracts every other particle through a force called *gravity.* It further gives a quantitative measure of how this attraction is larger the more massive the particles (doubling the mass of one particle doubles the force), and how it decreases when the distance between the particles is increased (doubling the distance weakens the force by a factor of four). To give another example of a law of physics, one of James Clerk Maxwell's equations, the laws that describe all the electric and magnetic phenomena, states that there are no magnetic monopoles (single magnetic poles). Namely, there cannot be a magnet that has only one, say, north pole. Indeed, we know that even if we take a bar magnet and chop it up into smaller and smaller pieces, each piece will have north and south poles.

Now, what is the meaning of symmetries of the laws of physics? These symmetries are certain fundamental properties of the laws, which are somewhat similar to the symmetries described for shapes or objects. For example, all the laws of physics do not change from place to place. A simple but remarkable manifestation of this property is the fact that if we perform an experiment, or study any physical phenomenon, in Russia, in Alabama, or on the moon, we obtain the *same* results. This also applies to different parts of the universe; when we observe a star that is located trillions of miles away from us, it still appears to obey the same laws of physics that we find here on Earth. This means that we can apply the same laws that we have deduced from laboratory experiments to the understanding of the universe as a whole. This universal transportability of the physical laws is encapsulated in the statement that the laws of physics are *symmetric under translations.* This property is not to be confused, for example, with the fact that the strength of the force of gravity is not the same on the earth and on the moon. Gravity on the moon is different (weaker) because both the mass and the size of the moon are different from those of the earth. However, given that we know the mass and radius of the moon, we would use *exactly the same formula* to calculate the force of gravity there, as we do on Earth.

The laws of physics also do not depend on the direction in space. For example, they would not change if the earth started to rotate in the opposite direction. Were this not the case, then experiments might yield different results in the Southern Hemisphere than they do in the Northern Hemisphere. Furthermore, we might obtain different results if we perform an experiment lying down, rather than standing up, or we might find that light travels faster to the north than it does to the east. Note again that I do not refer to the fact that, for example, different stars happen to be seen in the night's sky from Australia than from Alaska (nor to the fact that different rock music groups may be popular in the two places), but to the fact that the laws that describe all the natural phenomena do not have a preferred direction. Thus, the laws would not change if someone took our entire universe and rotated it somehow. This property is expressed by the statement that the laws of physics are *symmetric under rotation.*

I would like to further clarify the difference between a symmetry of a shape and of a law. The symmetry of the laws of physics under rotation *does not* mean, for example, as it was believed in ancient Greece, that the orbits of planets must be circular. A circle, as a *shape,* is indeed symmetric under rotation. But this has nothing to do with the symmetry of the *law*—in this case the law of gravity, which governs the motion of the planets around the sun. In fact, since the time of Johannes Kepler, a German astronomer who worked in Prague in the seventeenth century, astronomers have known that the orbits of the planets are not circular but elliptical. The symmetry of the law means that the orbit can have *any orientation in space* (Figure 3).

Another symmetry that the basic laws of physics exhibit concerns

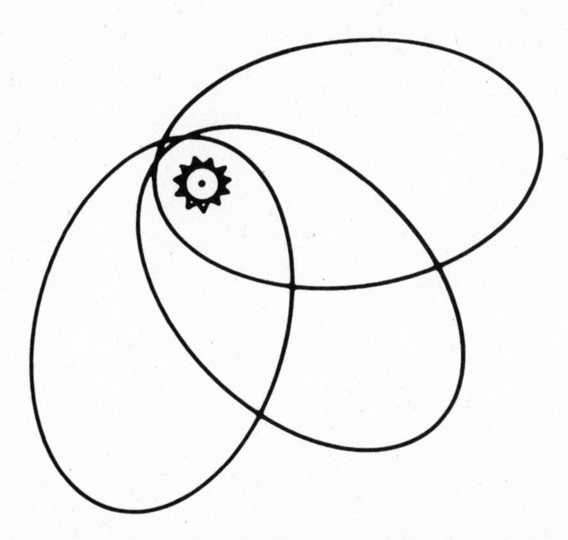

Figure 3

the *direction of time.* Curiously, the laws would not change if time were to flow backward.* This is true for both mechanical and electromagnetic phenomena at both the macroscopic and the subatomic levels. For example, there is nothing in the basic laws to indicate that the phenomenon of a plate falling from a shelf and shattering to pieces on the floor should be allowed, while that of the scattered pieces flying up from the floor and assembling to an intact plate on the shelf should be forbidden.

Interestingly enough, as far as we know, the laws of physics also do not change *with the passing of time.* Astronomy proves very useful in demonstrating this property. We can observe, for example, galaxies at distances of millions, and even billions, of light-years. One light-year is the distance light can travel in one year, which is about six trillion miles. What this means is that the light that reaches us now from a galaxy that is over 100 million light-years away *left that galaxy more than 100 million years ago.* Therefore, what we see today is really the way the galaxy was 100 million years ago. Astronomy thus truly allows us to look back into the past. The main point to note is that by analyzing the observed light, we can establish the fact that the same physical laws that govern the emission of light by atoms today also applied in the distant past. In fact, we can now state with a high degree of confidence that the laws of physics have not changed at least since the time the universe was only about one second old (see chapter 3).

As we have seen, when objects or shapes possess a certain symmetry, this is related to something that does not change, an *invariant.* For example, because of its left-right symmetry, a mirror image of the Notre Dame Cathedral in Paris looks identical to the

---

*Some readers may be familiar with the existence of the second law of thermodynamics, which is not symmetric with respect to the direction of time. However, as we shall see later, this represents in some sense not a fundamental law of physics but rather a dependence on the initial conditions. I shall return to the time symmetry in chapter 3.

cathedral itself. Symmetries are therefore related to the indiscernibility of differences. Since we are discussing symmetries of the laws that govern the behavior of all natural phenomena, the property of having things that do not change is in this case translated to universal entities we call *conservation laws.* A conservation law simply reflects the fact that there exist physical quantities in the universe that remain constant in time. Namely, if we were to measure the value of such a quantity today, one year from now, or a million years from now, we would obtain exactly the same value. This is to be contrasted, for example, with the stock market, where money is definitely not a conserved quantity—that is, on a given day everybody may lose, with no one gaining.

The two symmetries of the laws of nature I have already mentioned, the symmetry under translations and the symmetry under rotations, indeed result in conservation laws. For example, the *linear momentum* of a body is equal to the product of its mass and its speed, and its direction is the direction of the motion. Thus, the linear momentum of a body defines in some sense the quantity of motion this body possesses; it is larger the larger the mass and the speed. A stampeding buffalo has a larger linear momentum than that of a man running at the same speed but a smaller one than that of a rocket moving much faster. The symmetry under translations is manifested in the fact that linear momentum is conserved. Namely, momentum can neither disappear nor be created; it can just be transferred from one body to another. In everyday life, we see directly the consequences of conservation of linear momentum—for example, in the trajectories of colliding cars, of colliding billiard balls, and in the motion of the puck in ice hockey. The speed and direction of all of these motions are determined in such a way that the total momentum of the system is conserved. The motion of rockets is also a consequence of the conservation of linear momentum. When the rocket is resting on the launching pad, its momentum is zero (because its speed is zero). This means that as long as external forces do not interfere, the momentum must remain zero. When the rocket starts to eject gases downward at a high speed, the rocket itself acquires an upward speed, to counterbalance the momentum of the gases.

The *angular momentum* of a rotating body is a measure of the amount of rotation it possesses. For example, if two identical spheres are rotating around their axes, the angular momentum is larger for the one that rotates faster. If two spheres of the same mass are rotating at the same rate, the angular momentum is larger for the one with the larger radius. The symmetry of the laws of nature under rotations is manifested in the fact that angular momentum is also a conserved quality.

Ice skaters make frequent use of the conservation of angular momentum. In one of the popular routines, skaters start spinning slowly with their arms stretched sideways, and then they bring their arms to the sides of their body, thus increasing dramatically the rate of their spin. I still have a picture in my mind of a young Scott Hamilton, with his red hair spread almost horizontally, as he spins incredibly fast. This behavior results from conservation of angular momentum—reducing the distance of the arms from the rotation axis results in an increase in the speed of rotation. Conservation of angular momentum is also responsible (among other things) for the fact that moving bicycles and spinning Hannukah dreidels do not fall, for the stability of the axis of gyroscopes (which are used to determine directions accurately), and for the stability of the orbits of the planets around the sun.

Another symmetry that was mentioned above, the fact that the laws of nature do not change with the passing of time, is responsible for the existence and conservation of the quantity we call *energy*. We all have a certain intuitive understanding of what energy means; after all, we pay energy bills to gas and electric companies, and many of us still remember the energy crisis in 1979, when gasoline was expensive and hard to find. In some sense, energy reflects the ability to do work. Very broadly speaking, energy can be associated with motion (in which case it is called *kinetic energy*), can be stored in some form (e.g., chemical, electrical, gravitational, nuclear; in which case it is called *potential energy*), or can be carried by light (*radiative energy*). Again, conservation means that energy is neither created nor destroyed. It can merely be transferred from place to place or be transformed from one form to another. For example, when we drop a spoon, gravitational potential energy is transformed into kinetic energy of motion, and the latter is transformed into heat and acoustic energy as the spoon hits the floor.

Having briefly explained the concept of symmetry, I will now turn to the second requirement for beauty, that of simplicity.

2. Simplicity: Less Is More Beautiful

Simplicity is to be understood in the sense of *reductionism.* Namely, the goal of physics is to replace many questions by very few, basic questions; or a description of nature that involves many laws of physics by a complete theory that has only a few fundamental laws. Physicists have been driven for centuries by a feeling that underneath the enormous wealth of phenomena that we observe, there exists an underlying relatively simple picture. The great seventeenth-century French philosopher and scientist René Descartes once said: "Method is necessary for discovering the truths of nature. By method, I mean rules so clear and simple that anyone who uses them carefully will never mistake the false for the true, and will waste no mental effort." We have already seen an example of the application of this type of thinking in chapter 1, in the search for *one* mechanism to explain *all* the shapes of the nebulae.

We can identify in this drive for reductionism some of the same elements that perhaps formed the basis for the notion (in the Judeo-Christian cultures at least) that monotheism represents a more advanced (more beautiful?) form of faith than polytheism. The order for one God is expressed very clearly in the first two commandments: "I am the LORD your God . . ." and "You shall not make for yourself an idol . . ." I remember that as a skeptical child,

I used to be somewhat puzzled by the statement made by a teacher that the move to monotheism represented progress. After all, I thought, if it is all a matter of faith anyhow, then what difference does it make if you believe in one God or in many gods, each of whom is responsible for a different phenomenon in nature? Today, I can identify in that statement the same requirement for reductionism, for simplicity.

Given two theories that explain a given phenomenon equally well, the physicist will always prefer the simpler one, for this aesthetic (and not just practical) reason. I want to emphasize that this drive toward reductionism does not mean that the physicist fails to recognize that there is beauty in the richness and complexity of phenomena. After all, physicists realize, too, as did the poet William Cowper in the eighteenth century, that "variety's the very spice of life." The emergence of complexity in our universe, with life being perhaps at the pinnacle of this complexity, is what makes it so beautiful. However, in evaluating the beauty of a *physical theory,* the physicist regards as an essential element of beauty the fact that all of this complexity stems from a limited number of physical laws.

I would like to further clarify this idea with a simple example. Imagine that we draw a square, and then on each side of the square we draw another square with a side equal in length to one-third the side of the original square, and we repeat this process many times (Figure 4). Now, almost everyone will agree that the final pattern is quite beautiful to the eye (because of its symmetry). However, the physicist will recognize an additional element of beauty in the fact that underlying this relatively complex pattern, there is a very simple law (algorithm) for its generation.

The great German philosopher Immanuel Kant had similar ideas (in the eighteenth century) concerning the ideal of human consciousness. He defined this ideal as the attempt to establish our understanding of the universe on a small number of principles, from which an infinity of phenomena emerge. He went on to identify a beautiful object as one that has a multitude of constituents, all of

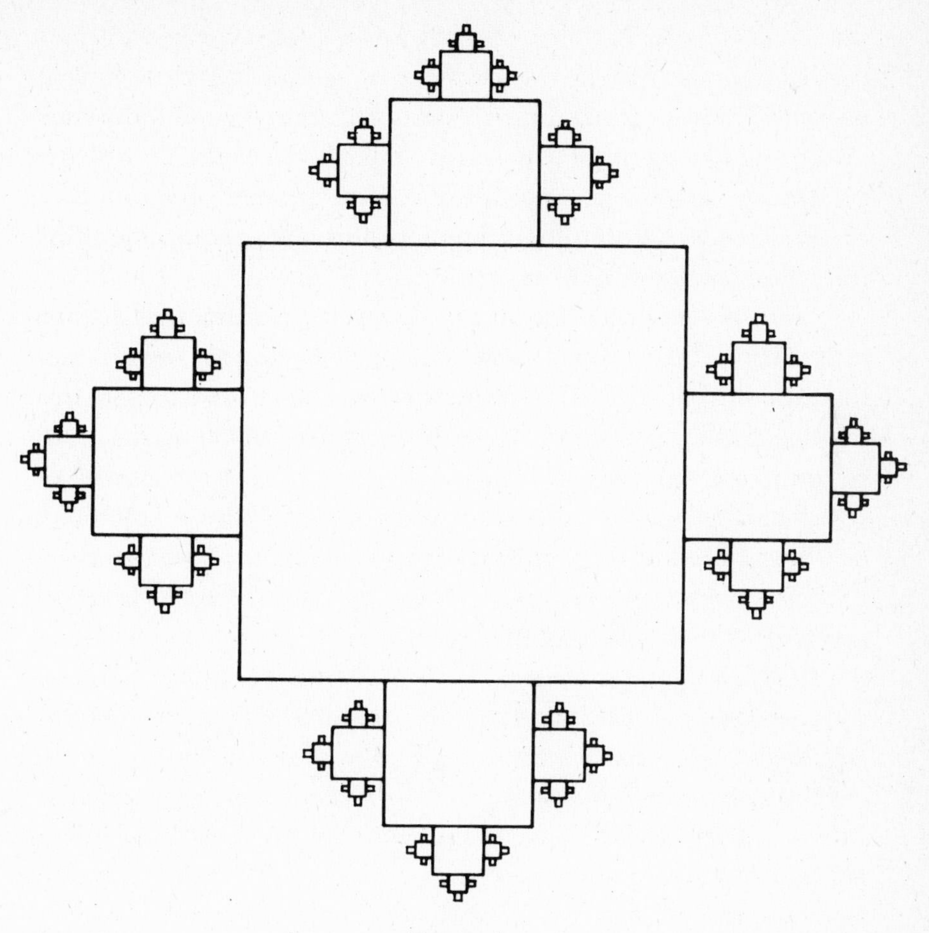

Figure 4

which at the same time obey a clear, transparent structure that provides the big picture.

Quite frequently, the mutual influences between the arts and the sciences are exaggerated. As a physicist who also happens to be an art fanatic, I can testify that the direct, immediate, conscious influence is minimal. Nevertheless, it is true that in some epochs, people from different disciplines sometimes think along similar lines. For example, a part of the title of this section, "Less Is More," was a popular aphorism with the twentieth-century architect Ludwig Mies van der Rohe. The feeling that one has to search for the most fundamental characteristics of things, which has guided physicists during all ages and in particular in this century, found its way also into some of the art movements of this century. Specifically, the roots of minimal art and conceptual art are clearly in this type of feeling.

An excellent example for this revolution in art is provided by the transition from the very realistic description of erotic attraction and the act of love, as in the sculptures *Desire* by the French artist Aristide Maillol or *The Kiss* of Auguste Rodin, to the very minimalistic description, as in the sculpture *The Kiss* by the French (Romanian-born) sculptor Constantin Brancusi. The entire train of thought (from realism to minimalism) is exposed in a series of tree paintings by the Dutch painter Piet Mondrian, in which one can literally follow the transformation of a very realistic rendition of a tree into a very abstract, minimalistic painting that involves a series of lines symbolizing the leaves.

Another painting by Mondrian that encapsulates the idea of reductionism, of grasping the most basic characteristics, is *Broadway Boogie Woogie* (currently in the Museum of Modern Art in New York City), in which the essence of the title is captured in a collection of squares and rectangles in bright, almost luminous, neon-like red, yellow, and blue colors.

Interestingly, while this reductionistic approach was adopted only by certain movements in Western art, by contrast, in Japanese art, as in physics, reductionism has been regarded as an element of beauty for centuries. It suffices to look at a lyric landscape painting by Toyo Sesshu from the fifteenth century, or to read a poem by Shikibu from the eleventh century

Come quickly—as soon as
these blossoms open
they fall.
This world exists
as a sheen of dew on flowers

to realize that the Japanese culture has brought simplicity and reductionism to aesthetic peaks. In fact, ever since the eighth century, the most popular poem structure in Japan has been the short poem (tanka), which has only five lines and thirty-one syllables (arranged in 5, 7, 5, 7, 7). In the seventeenth century, an even shorter structure appeared (haiku), with three lines and seventeen syllables (in a 5, 7, 5 pattern).

I will now explain the third element that I regard as essential for a physical theory of the universe to be beautiful—the *Copernican principle.*

3. The Copernican Principle: We Are Nothing Special

Many recognize Nicolaus Copernicus as the Polish astronomer who lived in the sixteenth century and reasserted the theory (based on a suggestion by Aristarchus some eighteen hundred years earlier) that the earth revolves around the sun. However, Copernicus was in fact (although not intentionally) responsible for a much more profound revolution in human thinking. The early models of the universe followed religiously the ideas of Aristotle, and were all geocentric. Namely, they all assumed that the earth was at the center of the universe. The most detailed, and most successful, model along these lines (in terms of explaining the observed paths of the sun, the moon, and the known planets) was due to the Greek astronomer Ptolemy, who lived in the second century A.D. This model survived, amazingly enough, for nearly thirteen centuries. One can only assume that it was the withering of intellectual curiosity during the Dark Ages, combined with the dominance of the Catholic Church, which regarded Aristotle's teachings as entirely consistent with its own doctrines, that granted the Ptolemaic model its longevity. In fact, following the integration of Aristotle's teachings into Christian theology, which is credited to Saint Thomas Aquinas (in the thirteenth century), Aristotle achieved an almost reverential status. Copernicus was the first to point out clearly that *we do not occupy a privileged place in the universe.* He discovered that we are nothing special. This has evolved to become known as the *Copernican principle.* In retrospect it is very easy to understand why there should be a Copernican principle in relation to the existence of "intelligent" creatures. After all, of all the places in the universe where intelligent creatures could emerge, assuming that there are many such places, very few, *by definition,* are "special." Therefore it is infinitely more likely for us to find ourselves in a nonspecial rather than in a special place. Put differently, the Copernican principle is a principle of mediocrity.

Since the time of Copernicus, the Copernican principle has been substantiated even further. Not only has the earth been dethroned from its central position in the universe, in fact, at the beginning of this century, the astronomer Harlow Shapley demonstrated that our entire solar system is not even at the center of our own Milky Way galaxy. Indeed, it is about two-thirds of the way out from the center, completing a revolution around the center in about 200 million years. As we shall see in the next chapters, this vulgarization of the earth's location continued even much further.

The Copernican principle can be expanded and generalized to include theories of the universe in general. In other words, every time that a certain theory would require humans to occupy a *very special place or time* for it to work, we could say that it does not obey the generalized Copernican principle. To give a specific example, if a theory were suggested in which the origin and evolution of humans was entirely different from that of all the other species, such a theory would not have obeyed the generalized Copernican principle. Darwin's theory of the origin of species by means of natural selection is thus a perfect example of a theory that *does* obey the Copernican principle (I will always mean the generalized principle from here on) and is therefore, from this point of view, beautiful.

I would also like to note that some theories are considered "ugly" because they violate something that can be regarded as intermediate between simplicity and an even more general interpretation of the Copernican principle. I include here all the theories that are extremely contrived, or that necessitate some very special circumstances or *fine-tuning* for their validity (even if they do not involve an explicit role for humans). The reluctance to associate beauty with such theories is a bit like the disbelief that we would surely feel if someone told us that he can flip a coin and make it land on its side. We will encounter examples for such fine-tuning in chapters 5 and 6.

I will now explain the concept of elegance, which, as I said, I personally do not regard as a *necessary* ingredient for beauty in a physical theory, but which can certainly enhance the beauty of certain theories.

4. Elegance: Expect the Unexpected

In mathematics and physics, and indeed in almost any discipline, it sometimes happens that a very simple, unexpected new idea resolves an otherwise relatively difficult problem. Such brilliant shortcuts lead to what are considered to be very *elegant* solutions. Interestingly, in chess, prizes for beauty are given for precisely this type of exceptional quality. It is amusing to note that one of the most elegant games of the brilliant American chess player Paul Morphy was played in Paris in 1858, in a box at the Paris Opéra, while Rossini's *Barber of Seville* was being performed on stage!

It is important to understand that elegance has nothing to do with reductionism (what I called simplicity). For example, the Ptolemaic model for the motion of the planets offered in fact an elegant solution to a difficult observational problem, in that it found a clever way to explain remarkably well the observed motions of the planets. The model, however, was not simple at all. It required each planet to move around a small circle, called epicycle, the center of which moved around the earth on a large circle (called deferent). To explain all the observations, the Ptolemaic model required no fewer than eighty circles! It was only following Kepler's discovery that the planetary orbits around the sun are elliptical that a *simple* model for the solar system emerged.

An example of elegance can be found in the following, well-known mathematical puzzle. Suppose we are asked: can one cover the board shown in Figure 5 by dominoes (each one having the area of exactly two squares), so that only one corner is left uncovered? The answer is very simple: no. Since the board has an even number (sixty-four exactly) of squares, and since each domino covers two squares, we can only cover an even number of squares and therefore we cannot leave one square open. Suppose, however, that we are now asked: can we cover the board in such a way that we leave two diagonally opposite corners uncovered? Clearly, in this case we will be covering an even number of squares, and so it is less trivial to determine immediately if this can be done or not (try thinking about this a little). It turns out, however, that with the help of an extremely simple trick we can answer the question right away. The idea is to think of the board as if half of the squares are painted black, as in a chessboard. Now, since each domino piece covers one

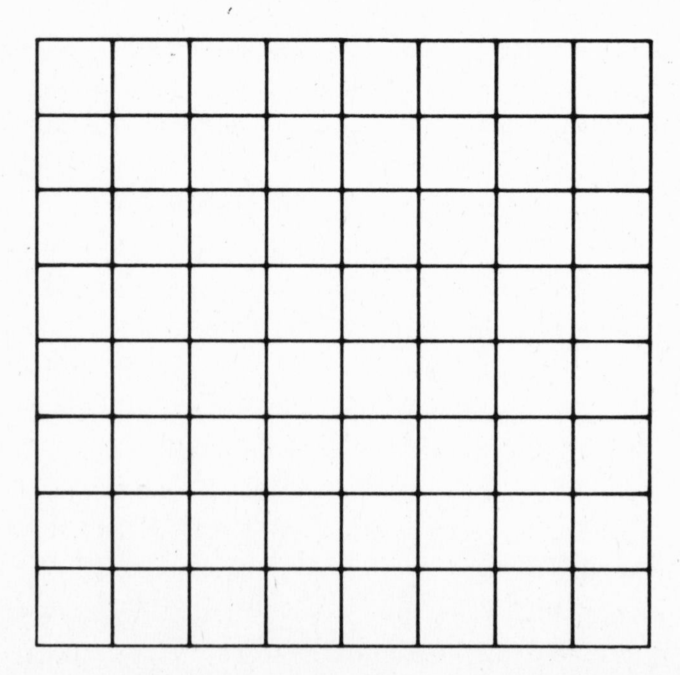

Figure 5

black square and one white square, it is clear that it is impossible to leave uncovered two diagonally opposite squares, since they have the same color! So, the extremely elegant idea of turning the board mentally into a chessboard helped us to solve immediately a problem that appeared otherwise to be much more complex. Isn't this elegant? However, I want to emphasize again that elegance, which I consider superfluous in a beautiful theory, should not be confused with reductionism, which I regard as absolutely essential.

She Walks in Beauty

The beauty (or absence thereof) of physical theories is the main theme of this book. Beauty, like love, or hatred, is also almost impossible to define properly. In fact, probably any definition is likely to raise some objections. I therefore feel that it is worthwhile, even at the risk of some repetitiveness, to attempt to reconstruct briefly the thought process that led me to my requirements. I should first note that because of the fundamental role I think beauty does play in physics, one cannot be satisfied merely with the attitude "I'll know it when I see it," which some physicists have adopted toward beauty in physical theories. I claimed to have identified three elements that are absolutely necessary for a physical theory to be beautiful. These are symmetry, simplicity, and a generalized Copernican principle. It is perfectly legitimate to question this identification and ask: (1) what is it that makes these elements essential to beauty? and (2) are there other elements that may be equally important?

The answers to these questions are not trivial, partly because of the fact that the recognition of beauty in a physical theory is to a large extent dependent on the scientist's intuition and sometimes even on taste (not unlike the dependence on the artist's aesthetic sensibility and taste in relation to a work of art). Nevertheless, I will attempt to outline now a certain logical process that can at least give us some guidance toward an answer. Instead of starting with the elements and trying to justify them, let us try to work our way backward, starting from a beautiful theory and retracing our steps in the direction of its basic ingredients.

First, it is important to realize that since the ultimate goal of physical theories is to describe the universe and all phenomena within it *in as perfect a way as possible,* a theory cannot produce a real sensation of beauty unless it can be regarded as a major step toward *perfection.* We therefore need to identify which combination of properties can be regarded as constituting perfection. Since the universe involves an immense number of phenomena, and may generally appear quite chaotic, it is clear that what is required is the introduction of some regularity, organization, balance, and correspondence into the description of nature. These properties can allow more encompassing perceptions, which eventually enable scientists to identify common characteristics of different phenomena. Next, since we are interested in beauty, we want to identify classical aesthetic constituents or concepts that can contribute. Adopting ideas from the arts, the list of such concepts may include symmetry, simplicity, order, coherence, unity, elegance, and harmony. The question now is which of these truly plays a central role in science. In order to answer this last question I will attempt to rank-order these concepts in terms of their contribution to scientific thinking.

Symmetry definitely occupies the top position in this hierarchy, since it literally forms the foundation on which physical laws are built, as I explained in the last section (and as we shall see further in chapter 3).

Simplicity comes next, since it allows scientists to choose from among all the different possible hypotheses and ideas the most economical ones. Albert Einstein himself wrote once: "Our experience until now justifies our belief that nature is the realization of the simplest mathematical ideas that are reasonable."

I did not list order, coherence, harmony, and unity as separate elements that are essential for beauty since in physics these are not independent concepts. For example, what is meant by order is that similar physical circumstances should produce similar consequences. However, we have already seen that symmetry and simplicity achieve precisely this goal. Furthermore, in the next section we will encounter an excellent example of the unifying power of symmetry and simplicity. Werner Heisenberg, one of the founding figures of quantum mechanics, the theory of the subatomic world, stated once as a criterion: "Beauty is the proper conformity of the parts to one another and to the whole." As we shall soon see, this is what symmetry and simplicity (reductionism) are all about.

Finally, as I discussed in the last section, elegance can certainly *contribute* to the feeling of attractiveness that is associated with a certain *solution* to a scientific problem. However, I do not regard it as an essential element of beauty in a physical *theory*. In this assertion I humbly disagree somewhat with the sixteenth-century philosopher and statesman Francis Bacon, who was described by the poet Alexander Pope as "the brightest, wisest, meanest of mankind." Bacon wrote that "there is no excellent beauty that has not some strangeness in the proportion." Since "strangeness in the proportion" is to be understood at least partly as an element of surprise, Bacon's "definition" is more closely associated with what I called elegance than with beauty. However, Bacon's criterion refers also to the unification of otherwise seemingly independent concepts, and as such to symmetry and reductionism. My point of view on elegance is best expressed by a famous quote from the nuclear physicist Leo Szilard: "Elegance is for tailors."

I hope that the above discussion clarifies the first two elements in my definition of beauty in physics. The third element, however, requires some further explanation. I have included in my essential ingredients the generalized Copernican principle, which is not traditionally linked with aesthetics. This is really a property that is peculiar to the sciences, and to theories of the universe in particular. In general, scientists absolutely detest theories that require special circumstances, contrived modeling, or fine-tuning. In this sense, as particle physicist Steven Weinberg puts it in his book *Dreams of a Final Theory*, a beautiful theory must be seen as essentially *inevitable.*

A violation of the generalized Copernican principle, in the form of statements like "the universe must be so-and-so because we humans are so-and-so," or in the form of fine-tuning, is certainly a slap in the face of all encompassing inevitability, and is therefore ugly. I will return to this question in the discussion of intelligent life and the anthropic principle in chapter 9. At this point I will conclude by noting that placing humans center stage may be regarded as a desirable property from theological, psychological, or even theatrical perspectives, but it is not a property most scientists would like to associate with a beautiful theory.

I do not wish to leave the reader with the impression that in their quest for beauty in the theory physicists lose sight of the beauty of the universe itself. This is certainly not the case. When Einstein once said that the only incomprehensible thing about the universe is that it is comprehensible, he referred precisely to these two "beauties." The reality of Marc Chagall's *Lovers with Half-Moon* (currently at the Stedelijk Museum, in Amsterdam) is much more than the chemical composition of its paint. Humans are on one hand able to access the beauty of the universe and on the other able to comprehend the beauty of its workings.

I have always admired the theater. I regard the art of presentation of ideas succinctly through dialogues and monologues as being a close cousin to the "art" of scientific presentations. Consequently, I have decided that in a few places in this book, I will abandon the more expository style in favor of a more theatrical style. The aim of these short "scenes" is to provide an introduction to new concepts, on the basis of the ideas that have already been developed.

Galileo

In a dark room, lit only by a few candles, five men dressed in red sit behind a long table. The grave expressions on their faces make them look almost identical.

MAN AT THE CENTER OF THE TABLE (GRAND INQUISITOR): Bring him in! [*The steps of a guard echo from the stone walls and the ceiling. The guard enters, pushing in front of him a bearded old man, who clearly has trouble walking.*]

GRAND INQUISITOR: We would like to ask you again a few questions about your crazy and, may I add, dangerous ideas.

OLD MAN: All of my ideas merely represent the progress of scientific knowledge.

GRAND INQUISITOR: We will be the judges of that. The proposition that the sun is in the center and immovable from its place is absurd, philosophically false, and heretical. Do you agree that the earth stands still at the center of the universe?

OLD MAN [*in a weak voice*]: Certainly not. The earth rotates around its axis and at the same time revolves around the sun. The entire solar system is not even at the center of our own galaxy.

INQUISITOR at far left end of the table [*with surprise*]: Galaxy? What galaxy? What does the Milky Way have to do with it?

OLD MAN [*straightening himself up a bit]:* While the faintly luminous band seen across the sky at night is referred to as the Milky Way, a galaxy is really a collection of about one hundred billion stars like our sun.

GRAND INQUISITOR: Are you out of your mind? Where are all these suns?

OLD MAN: As I said, our own solar system belongs to such a galaxy. Most of these stars are too faint to be observed with the naked eye.

GRAND INQUISITOR: Why do I not see that the earth revolves around the sun as you say? I see everything here standing still, while the sun is moving around the earth.

OLD MAN [*somewhat scornfully*]: That is because you can only see relative motions. Everything on the earth is moving with the earth, so you do not see it moving with respect to you.

SECOND INQUISITOR FROM RIGHT: This is the greatest nonsense I have ever heard, even if I were to ignore the desecration in your words. Soon you will be telling us that the sun is not revolving around the earth but around something else.

OLD MAN: Indeed, the sun is rotating around its own axis, and it is revolving around the center of our galaxy.

GRAND INQUISITOR [*raising his voice in great anger*]: Do your ears hear what your mouth utters? Everything is rotating! [*With great scorn.*] What else do you think is rotating?

OLD MAN [*now clearly hesitant*]: Well, the electrons in the atom, for example, revolve around the nucleus.

[*The inquisitors are visibly baffled by this statement and start whispering among themselves. Finally, the Grand Inquisitor resumes the interrogation.*]

GRAND INQUISITOR: It is becoming more and more obvious to us that you have lost your faculties. While we do not have the faintest idea of what you are talking about, I point out to you that the word *atom* in Greek implies that it is indivisible and therefore [*raising his voice*] there cannot be anything in it!

[*The old man remains silent.*]

GRAND INQUISITOR: Well?

OLD MAN [*feebly*]: Matter as we know it is made of atoms, this is true. But the atoms themselves have a very dense and compact nucleus at their centers. In this nucleus there are particles called protons and neutrons. Other tiny particles, called electrons, revolve around this nucleus.

GRAND INQUISITOR [*clasping his hands and looking upward in despair, then, after looking at his colleagues, turning to the old man with a resentful face*]: At least I hope that your protons and electrons are not rotating around their axes?

[*All the inquisitors laugh loudly.*]

OLD MAN [*with some determination*]: Well, the electrons and the protons have a quantum mechanical property called "spin," which in some respects can be thought of as if they are rotating around their own axes.

GRAND INQUISITOR [*shouts in rage*]: Shut up! I've had enough of this. The earth does not revolve, it is at the center of the universe, and all of these so called electrons and protons do not even exist. [*Turning now to the guard.*] Put him in a damp cell in solitary confinement, where he will have time to think until his own head will start spinning!

[*The guard starts to drag the old man out of the room. The old man, his eyes open wide with fear, can hardly keep up with the guard's huge steps. As he is being dragged past the large wooden door he murmurs to himself]:* Eppur si muove. (And yet it moves.)

Needless to say that in Galileo Galilei's time (1564–1642) galaxies, atoms, nuclei, quantum mechanics, and electrons had not yet been discovered. But had they been, it would not surprise me if Galileo would have used about them similar phrases to the ones I have put in his mouth.

Spinning

Electrons are the smallest (in mass) known particles that are electrically charged. Galaxies are huge collections of tens of billions stars. Yet precisely the same physical laws govern the behaviors of both. This is the true meaning of simplicity and symmetry—of beauty in physics. How do we know this? The following example describes these principles in action.

Electrons and protons, the basic building blocks of atoms, have a property called *spin.* Strictly speaking, this property can be described only by quantum mechanics—the theory that governs the subatomic world. However, for some purposes spin can be thought of as a rotation of the electron or the proton around its own axis. Therefore, as for all rotating bodies, the properties of this spin are described by the conservation of angular momentum. Because of the fact that electrons and protons also have an electric charge, this spin makes them behave like small bar magnets, since a magnetic field is generated when electric charges are moving. Now, a simple experiment can be performed with two small magnets, such as two compass needles. Imagine that you suspend the needles on thin threads attached to their middle points. If you place the two needles along a straight line, with two *equal* (say, north) poles facing one another, you will observe that one of the needles will spontaneously flip, so that two *opposite* (one north and one south) poles will face one another. Energetically speaking, when a system undergoes such a spontaneous transition, this means that the second configuration is at a lower energy state than the first, since physical systems tend to be in their lowest possible energy state (as I am sure everybody recognizes from their own tendencies).

In an analogous manner, when the hydrogen atom, which is composed of one proton around which one electron is orbiting, is in its lowest orbital energy level, the spins of the electron and proton can be parallel, which is equivalent in some sense to the electron and proton rotating in the same sense, or antiparallel (namely, the electron and proton rotating in opposite senses; quantum mechanics only allows two possible states for the spins of the electron and the proton). Consequently, the hydrogen atom can spontaneously undergo a *spin-flip transition,* from the parallel to the antiparallel state. Since the latter is at a lower energy state, the difference in the energy is emitted in the form of radio waves. Waves in general are characterized by a property called *wavelength.* For example, when we throw a pebble into a pool, we can observe a series of concentric waves, and the distance between two crests is the wavelength.

The wavelength of the radio waves emitted by the spin-flip transition is 21 centimeters. So, how is all of this connected to galaxies? As it turns out, this 21-centimeter radio wave plays a crucial role in our exploration of galaxies. For example, in order to determine the global structure of our own galaxy, the Milky Way, and the motions within it, observations must be made to distances of the order of 30,000 light-years. The problem is, however, that the interstellar medium, diffuse gas and dust in which clouds are dispersed, prohibits optical observations to such large distances because it is opaque to visible light. The idea is therefore to identify a wavelength of electromagnetic radiation to which the interstellar matter is relatively transparent, so that the source of the radiation can be seen to large distances. In addition, the wavelength has to be such that either stars or cold gas clouds, which are the main and most abundant participants in the motions in the galaxy, would be strong emitters in that wavelength, to allow its detection.

The 21-centimeter radiation proves perfect for this purpose. Not only does it allow observations to penetrate to the farthest corners of the Galaxy, it is also emitted by neutral hydrogen atoms, which constitute the main component of interstellar matter, in the form of cold gas clouds.

There is one more effect that needs explanation in relation to observations aimed at determining the speeds of gas clouds, and this is known as the Doppler effect, after the Austrian physicist Christian Doppler, who identified it in 1842. We are familiar with this effect in everyday life, in relation to sound waves. When a source of sound (e.g., a car or a train) is approaching us, the sound waves bunch up and we receive waves at a higher frequency (higher pitch). The opposite happens when the source of sound is receding from us. The waves are spread out and we hear a lower pitch. The effect is particularly noticeable when the source of the sound passes us by rapidly, since the sound changes from a high to a low pitch (recall, for example, the "eeeeeeoooooo" sounds of cars on the Indy 500 race track). A similar phenomenon occurs with light or any electromagnetic radiation. Therefore, when a source that emits the 21-centimeter radiation is receding from us, the radiation will be detected at a *longer wavelength* (lower frequency), while when it is approaching us, it will be observed at a *shorter wavelength* (higher frequency). From the detected shift in the wavelength, the speed of the source can be determined, since the shift is larger the higher the speed.

The stars and the gas clouds in the Galaxy participate in a global rotation around the galactic center. Radio observations of the 21-centimeter radiation allowed astronomers to determine the structure and the rotation pattern of the entire galaxy interior to the sun's orbit—namely, between the solar system and the center of the Galaxy.

A detailed mapping of our galaxy reveals that it is a disklike, spiral galaxy. Namely, it has the shape of a flattened pancake, with a spiral pattern observed on the face of the disk. The disk structure is probably a consequence of the formation process of the Galaxy, and it is related at least in part to the amount of angular momentum (amount of rotation) possessed by the gas cloud from which the Galaxy formed. The exact process by which galaxies form is still not fully understood; however, the following general remarks can be made. When a rapidly rotating cloud of gas collapses to form a galaxy (due to the force of gravity), it will tend to form a flattened disk. This is a consequence of the centrifugal force, which tends to push material away from the rotation axis. The centrifugal force is very familiar to anyone who has tried to negotiate a sharp turn with a car moving at a high speed—a strong force is felt, pushing the car off the road. The centrifugal force is stronger when the "amount of rotation" of the cloud is higher. Consequently, for high angular momentum clouds, the collapse forms a flat, disklike structure.

So, what have we have discovered here? This same entity, called angular momentum, the conservation of which resulted from the symmetry of the laws of physics under rotation, explained to us something about the *electron,* which has dimensions of about 10^{-13} (0.0000000000001) centimeter, and about the *Galaxy,* which has dimensions of about 10^{23} (100,000,000,000,000,000,000,000) centimeters. What's more, *we used one* (the spin-flip of the electron) *to discover the other* (the structure of the Galaxy). Now, I am asking you, isn't this absolutely BEAUTIFUL?

3

Expansion

Modern cosmology is the study of the origin and the evolution of the universe as a whole. It deals with such fundamental questions as: What is the origin of the presently observed expansion of the universe? What is the origin of matter? What were the seeds from which the structures that we observe in the universe today, such as galaxies, emerged? These are not easy questions. The mere fact that we believe that we are coming close to being able to answer these and similar questions can in itself be regarded as absolutely amazing, given the fact that only a hundred years ago many of the questions did not even exist!

Since the laws of nature have the elements of beauty engraved in them, it should come as no surprise that aesthetic principles played a major role in the shaping of our thinking about the origin of the universe.

I will start with a simple, apparently unrelated example, which will turn out to have an enormous significance.

Imagine that you are standing in the middle of a forest, in which all the tree trunks have been painted white. If the forest is not very large, then probably in some directions you can catch a glimpse of what is outside the forest through the gaps between the trees, since your line of sight in those directions does not intersect any trunk. Imagine now, however, that the forest around you is made thicker and thicker, simply by adding trees around the initial forest's periphery, to larger and larger distances. Clearly, at some point, it will appear to you that you are surrounded by an environment that is uniformly white, because in whichever direction you would look, your line of sight would intersect a tree trunk. As we shall soon see, this example is intimately related to some properties of the universe.

From the early days of modern cosmology, at the beginning of the twentieth century, the following basic assumptions were made: that the universe is *homogeneous*—namely, it has the same properties in every part of it—and *isotropic*—namely, it looks the same in any direction. Together, these two basic foundations are known as the *cosmological principle.* Notice that these two assumptions coincide precisely with the ingredients of my definition of beauty, since homogeneity is the simplest assumption one can make (and it also implies symmetry under translations), and isotropy implies that there is no preferred direction and therefore a symmetry of the universe under rotations. Furthermore, homogeneity and isotropy also reflect the idea behind the Copernican principle, since the fact that we are not center stage in the cosmos immediately implies that our location is typical, and therefore, looked upon from any other location, the universe at large would look the same.

Interestingly, the idea that the universe is homogeneous dates back to the sixteenth century. The Italian philosopher and scientist Giordano Bruno asserts in his book *On the Infinite Universe and Worlds* (published in 1584) that the universe has the same structure throughout, consisting essentially of an infinite repetition of solar systems (in his words, "*synodi ex mundis,*" or "assemblies of worlds"). Unfortunately, Bruno's enthusiastic championship of Copernicus (and really of the idea behind the Copernican principle) brought him into direct confrontation with the Inquisition. He was burned at the stake in Rome in 1600, after a seven-year trial.

Powerful modern telescopes on the ground and in space place our observational horizon at about ten billion light-years. Observations into these depths of space reveal that homogeneity and isotropy are not merely *assumptions.* As far as we can tell, the universe can indeed be described by a broad-brush uniformity. It appears that on the largest scales (larger than a few hundred million light-years) the universe actually *is* homogeneous and isotropic.

These two observed properties (homogeneity and isotropy) have two important consequences: The universe has no *edge* and no *center.* If the universe had an edge, then clearly all the places within it would not have been the same. For example, a place near the edge would have been quite different in terms of its environment than one near the center (in the same way that a place near the north edge of Queens is not the same as one in the middle of Manhattan). This, however, would contradict the homogeneity. Similarly, if the universe had a center, things would not have looked the same in all directions from a point that was not precisely at the center, contradicting the isotropy.

Let us think for a minute of how is it possible for a geometrical configuration not to have an edge or a center. One obvious example is an infinite, homogeneous flat plane. By definition, it has no center or edge. However, the configuration does not have to be infinite not to have an edge. Consider for example the two-dimensional *surface* of a spherical balloon. Again, it has no center, because there is no special point on the surface of the balloon (note that the center of the balloon itself does not belong to the surface, which is the geometrical entity we are considering). The surface also has no edge; if we were two-dimensional creatures living on this surface (like dots painted on the balloon), and we were to travel continuously in a given direction, we would never reach an edge; in fact, we would eventually return to our starting point.

Now, imagine for a moment that in addition to being homogeneous and isotropic, our observable universe were also *infinite,* with *identical* stars scattered throughout, like raisins in a cake. In this case surely, because of the infinite dimensions, our line of sight in *any* given direction would eventually hit the surface of a star, just as in the example of the forest with the white trees. Consequently, we would expect to find the night sky to be *uniformly bright,* with the brightness of a star, just like the uniformly white surroundings in the forest. This expectation is in stark contrast to the real appearance of the night sky—dark, with stars appearing like points of light sprinkled in the darkness. This apparent paradox is known as Olbers's paradox, after the nineteenth-century German astronomer who phrased it in 1823, although the general problem had been known at least a hundred years earlier, and possibly since Johannes Kepler. Since the validity of the assumptions of homogeneity and isotropy on a large scale has been confirmed by modern astronomical surveys, this huge discrepancy between the prediction—*uniformly bright sky*—and the observation—*dark sky*—must mean that one (or both) of the assumptions—"the observable universe is *infinite*" and "with *identical* stars scattered throughout"—must be wrong. It should be noted that the paradox cannot be resolved by simply assuming that starlight is blocked by some interstellar dust, because in equilibrium such a dust would absorb the intense starlight and reradiate the energy just like a star, or would be evaporated away. As incredible as this may sound, it turns out that this simple fact, that the night sky is *dark,* with which we have become so accustomed that we hardly ever give it a second thought, derives from a property of the universe the discovery of which is probably the most dramatic in cosmology.

In the 1920s, the astronomer Edwin Hubble (after whom the Hubble Space Telescope is named) discovered the following observational facts (some of which had been known from previous work by the astronomer Vesto Slipher):

1. *Every* distant galaxy appears to be receding from us.
2. More distant galaxies are receding faster than closer galaxies, in such a way that a galaxy twice more distant recedes twice as fast.

The first question that arises in relation to these discoveries is: how does one discover something like this? The answer is: with the help of the Doppler effect. As I explained in chapter 2, when a source of light (or any type of waves) is receding from us, the light is observed to be shifted to a longer wavelength or lower frequency. Since in the visible (to the human eye) part of the electromagnetic radiation, the longest wavelength is red, it is said in such a case that the light is *redshifted.* Now, every atom emits radiation at a particular set of wavelengths, which corresponds to the differences between its different energy states. For example, when a hydrogen atom undergoes a transition to its lowest energy state from the energy state just above that, the difference in the energy is emitted in the form of ultraviolet radiation, with a wavelength of 0.00001216 centimeter. Every atom has its own unique set of energy levels and therefore its own set of wavelengths, which is called the *spectrum* of that atom. Therefore, when radiation that corresponds to a wavelength in the spectrum of a given atom is observed, this serves as a fingerprint indicating the presence of that atom. What Hubble discovered was that the spectra observed from all the galaxies were all redshifted, indicating recession. Since the amount of shift is determined by the speed of the source (the faster the source, the larger the shift), he was able to determine that the speed of the recession is proportional to the distance. Present-day telescopes are able to establish that some of the very distant galaxies, some of which are as distant as 10 billion light-years, and the objects known as *quasars* (the nuclei of galaxies that show enormous activity) are moving away at speeds exceeding 90 percent of the speed of light.

The second question that needs to be asked is: how is it possible that everything is receding from *us*? After all, who are *we* in the grand scheme of the universe? Does this not violate the Copernican principle, by implying that we are at the center of the universe? Why is it not the case that some galaxies are receding from us while others are approaching? One might almost think that we should be offended by such an expression of apparent cosmic antipathy toward us!

It is easiest to understand this observation in terms of an imaginary universe that has only two spatial dimensions. In fact, I have already used such a universe earlier in this chapter, when I explained the concepts of homogeneity and isotropy—the surface of a spherical balloon. As I noted then, this two-dimensional universe has no center and no edge. Imagine now that we are two-dimensional creatures living on this surface in some painted dot that represents a galaxy. Note that the fact that the balloon itself exists in some three-dimensional space is not something that is easily perceived by the inhabitants of this universe, since the third dimension does not exist for them. What will happen now if this balloon is expanding? In such a case, we will see that *all* the galaxies around us are receding from us. This observation will remain true irrespective of the galaxy on which we happen to reside. Furthermore, if we examine two galaxies, one of which was initially twice more distant than the other, the more distant galaxy will appear to be receding twice as fast. In other words, in this two-dimensional universe we will discover exactly what Hubble discovered. The essence of the finding remains the same in our three-dimensional space—namely, *Hubble discovered that our universe is expanding.*

Interestingly, Hubble himself was never fully convinced of the expansion. In a 1936 paper he wrote, "The high density suggests that the expanding models are a forced interpretation of data." Hubble's skepticism stemmed mostly from a parallel observational program in which he counted the number of galaxies at every brightness range. Underestimating his errors in this count analysis, he thought that the results he obtained were inconsistent with an expanding universe. Hubble's doubts about the reality of the expansion continued even up to his last paper in 1953.

Hubble's discovery of the expansion of the universe and the fact that the speed of recession of galaxies is proportional to their distance have some dramatic consequences. But first, it is important to emphasize again that according to the cosmological principle of homogeneity and isotropy, *the universe has no center.* An observer on *any* galaxy would see exactly the same expansion of the universe as any other observer on any other galaxy. Second, I would like to clarify that it is only the scale of the universe at large, as expressed by the distances that separate galaxies and clusters of galaxies, that is expanding. The galaxies themselves are not increasing in size, and neither are the solar systems, individual stars, or humans; space is simply unfolding between them. Furthermore, a common misunderstanding is to think of the galaxies as if they are *moving* through some preexisting space. This is not the case. Think of the dots on the surface of the balloon. Those dots are not moving at all on the surface (which is the only *space* that exists). Rather, *space itself* (the surface) is stretching, thus increasing the distances between galaxies. Finally on this point, the limit from special relativity that matter cannot move faster than light does not apply to the speeds at which galaxies are separated from each other by the stretching of space. As I explained above, the galaxies are not really moving, and there is no limit on the speed with which space can expand. As an aside, I should note that unlike in the case of moving cars or trains, the Doppler effect used to determine the redshift is also in this case a stretching of the wavelength due to the expansion of space itself. Third, and most important, Hubble's relation between distance and speed of recession tells us that *our observable universe had a beginning.* Our observable universe did not always exist; rather, it has an age of about 11 to 15 billion years (I will use 14 billion from here on). To see how we can conclude the latter, recall how we can estimate the time it would take us to drive a certain distance. For example, the distance from Baltimore to New York is about 210 miles; we know then that if we drive at an average speed of 60 miles per hour, it will take us three and a half hours to get there, because the time is equal to the distance divided by the average speed. Similarly, since Hubble's discovery relates the distance of each galaxy to its speed, we can determine approximately how long ago the expansion of the universe started (a more precise determination requires knowing also if there were periods of acceleration or deceleration). Note that it is *the entire universe* that is expanding; that is, every galaxy is moving away from every other galaxy. If we were to run time backward, like rewinding a video, all the galaxies would eventually collapse upon one another (if galaxies were to exist that early in the age of the universe; see below). In other words, as with the expanding spherical balloon, which if completely deflated would have shrunk to a minuscule size, our universe *started from one point.* Specifically, about 14 billion years ago, all the matter and energy that we can observe in the universe today were concentrated in a sphere smaller than a dime in radius.

Let us now return to Olbers's paradox—the fact that the night sky is dark, while we expected it to be as bright as the surface of a star. We discovered that this paradox implies that one (or both) of the assumptions that the observable universe is *infinite* and filled with *identical* sources of light (stars or galaxies) has to be wrong. Hubble's discovery of the expansion of the universe and the fact that the universe has a finite age of about 14 billion years help to resolve the paradox. Since the universe is 14 billion years old, it means that even in principle we cannot see stars or galaxies beyond a certain "horizon," because even if there are sources of light out there, the light from those sources *did not have enough time to reach us during the universe's lifetime.* This means that to our eyes the universe is not infinite (even if it were infinite in extent), making that assumption invalid. It would be as if in the example with the forest, from a certain distance on, all the tree trunks are actually black instead of white. Clearly, in this case we will not see a uniformly white environment. Second, because the universe is expanding, the light from every distant galaxy is redshifted to a lower frequency. Electromagnetic radiation is less energetic the lower its frequency. For example, photons of X rays are more energetic than photons of visible light because their frequency is higher (photons are the carriers of electromagnetic energy). Consequently, the more distant the galaxy, the less energetic the photons we receive from it (because it recedes faster, and is therefore more redshifted). This means that again, from our point of view, the assumption of *identical* sources of light also breaks down. In the example with the forest, this would be equivalent to having the trunks painted unevenly, with more distant trees being less and less painted.

Is it not absolutely amazing that the seemingly simple fact that the night sky is dark had to wait for the discovery of the expansion of the universe, and the realization that the universe had a beginning, to find its explanation? One can hardly think of a better demonstration of how what appear to be entirely unrelated phenomena become interwoven through the application of the same basic laws.

The fact that we see a dark, starry night is therefore the first major discovery in cosmology. It is interesting that long before humans realized this fact, starry nights already generated an incredible fascination and had an extraordinary emotional impact. In the biblical description of creation, the stars are created in the dome of the sky on the fourth day. Later in the book of Genesis, after Abraham is prepared to sacrifice his son Isaac, he hears the promise: "I will make your offsprings as numerous as the stars of heaven." Personally, I find the stars' appeal best expressed perhaps in two remarkable paintings entitled *Starry Night,* one by the Norwegian artist Edvard Munch and one by the Dutch painter Vincent van Gogh.

While Munch is justifiably regarded as one of the best modern presenters of the ancient Nordic spirit and van Gogh is traditionally labeled a Postimpressionist, there is no question that both of these artists can be considered among the founding figures of modern expressionism. Perhaps not accidentally, both painters were also hospitalized for mental illness. Van Gogh's *Starry Night* (now in the Museum of Modern Art, in New York City) was painted in June 1889, only one month after his admission, at his own request, to the asylum at Saint-Rémy and one year before his suicide. Munch's *Starry Night* (now in the J. Paul Getty Museum, in Malibu) was painted in 1893—in fact, sometime after he may have seen van Gogh's painting at the dealership of Vincent's brother, Theo, in Paris.

The paintings have several things in common, in spite of their very different appearances. For example, both painters painted the bright stars in their true positions in the night sky, for the time frame in which the paintings were executed (Munch actually included the planet Venus in his painting). The stars in both paintings are heavily ringed with halos of light. This is particularly true in van Gogh's painting, and in Munch's two paintings with the same title, painted in 1923–1924. Interestingly, the appearance of the stars in these paintings is remarkably similar to images of stars obtained with the infrared camera on board the Hubble Space Telescope. In these images, the ringed halos are a result of the diffraction patterns produced by the optics.

Munch's painting is almost abstract, and it conveys a troubling, mystical mood (a wonderful description of all the painting's history can be found in *Edvard Munch, Starry Night,* by Louise Lippincott). The view in the painting is seen from a window in the Grand Hotel in Åsgårdstrand, the same place where eight years earlier Munch started an affair with Milly Thaulow, his cousin's wife. In describing the mood, which eventually diffused into the painting, Munch wrote: "I am so fond of the darkness—I cannot stand the light—it ought to be just like this evening when the moon is behind the clouds—it is so mysterious . . ."

Van Gogh's *Starry Night* is a tumultuous painting, in which clear premonitions of sufferings to come are articulated. Nighttime scenes had just become interesting to van Gogh, because of the need to use visual memory. One account claims that the painting was inspired by a dream of traveling to the stars in death. The small village in the picture is framed with van Gogh's recurrent motifs in Saint-Rémy—a cypress, olive trees, and the undulating shape of the Alpilles. The general atmosphere is one in which the natural elements, in particular the dramatic stars, appear almost menacing (a detailed description appears, for example, in *Vincent van Gogh,* by Ingo Walthers and Rainer Metzger).

Overall, the two paintings offer a prime example of the emotional response of humans to the power and vitality of nature. It is impossible to imagine how the two painters could have made their respective starry nights even more charged, had they known that the latter represent such an incredible property of the universe as expansion.

Hubble's discovery of universal expansion also expands the horizons of the Copernican principle. It clearly shows that our galaxy is remarkably ordinary and not in the center of the universe. In fact, Hubble's observation indicates that, consistently with the cosmological principle, the universe does not even have a center. Curiously, this last conclusion led to an interesting incident in which I became involved recently. In July 1998, an astronomer colleague, Carol Christian, and I were invited to appear on the "Marc Steiner Show." This is a popular radio program on Maryland Public Radio in which the host, Marc Steiner, brings in guests to talk on a wide range of topics. We were invited to discuss recent astronomical discoveries. During the program I mentioned Hubble's discovery, and its meaning. The following day I received an e-mail from a listener who suggested the following interpretation of my comments. The message said: "Isn't it true that unlike Copernicus, who has put us in the center, Hubble actually showed that any point in the universe is the center?" The message continued by suggesting that this should have some deep theological implications. I answered that while I felt in no position to discuss the theological issues, Hubble's discovery and indeed the cosmological principle itself certainly imply that all locations in the universe are *equivalent.*

A Hot Start

The fact that the universe started from a point some 14 billion years ago and has been expanding ever since is known as the *hot big bang* model. In this model, the universe was not only extremely dense at the beginning but also extremely hot, having a temperature of about 10^{32} Fahrenheit (one followed by 32 zeros; actually, at such temperatures it does not really matter if one uses Fahrenheit or Celsius) at the time its dimensions were about 10^{-33} centimeters (0.00 . . . . . 1, with one at the thirty-third decimal place). As the universe expanded, it also cooled at a rapid rate, due to the shift of radiation to a lower frequency, which reduced its energy. By the time the temperature had dropped to about 100 million times that in the center of our sun (the latter being about 30 million degrees Fahrenheit), the basic building blocks of matter formed. These are the elementary particles called *quarks,* which combine to form the protons and neutrons of ordinary matter. After the universe had expanded 1,000 times more, the quarks became confined in neutrons and protons, and it took an additional expansion by a factor of 1,000 before the protons and neutrons started to combine to form nuclei of light atoms. In particular, most of the nuclei of the element helium that are present today, about 23 to 24 percent of all the visible matter in the universe, were formed in that era. Everything I have described so far occurred during a period of about a minute from the start of the expansion. The universe was still too hot, however, for electrons to be captured by nuclei and remain bound to them to form atoms—the electrons were too energetic. Under these conditions, radiation could not travel very far without encountering some particles, especially electrons, and being absorbed and reemitted or scattered by them. The universe was therefore opaque to electromagnetic radiation. The fact that energy was continuously being exchanged between the particles and the radiation led to a state in which the temperature was quite uniform, at least across small distances in the universe, since any change from the local average temperature could be immediately erased by exchange of energy. This balance between matter and energy is known as a state of *thermal equilibrium,* and it is characterized by the fact that the intensity of the radiation at each wavelength is determined uniquely by the temperature, to produce what is known as a *thermal spectrum.* Only after the universe was about 300,000 years old, and only about 1,000 times smaller than today, did it become cool enough for normal, neutral atoms to appear. At that point, since the electrons were captured inside atoms in orbits around the nuclei, radiation could travel essentially freely in the universe, without ever being absorbed or scattered (except for radiation at very special wavelengths, corresponding to the differences between the energy levels of the hydrogen and helium atoms). The situation is somewhat similar to that in my neighborhood's swimming pool. There are at any given moment a few parents and many children in the water. As long as the children are allowed to swim freely, it is impossible to swim more than a yard without bumping into some child. In the few instances, however, at which the parents decide to keep their children close to them, suddenly large gaps and open spaces for swimming develop. The universe therefore became transparent to radiation, which, unhindered, filled the entire space, forming a cosmic radiation background. Since this time in the "life" of the universe is marked by the decoupling of radiation from matter, it has become known as the *decoupling epoch.*

One of the most spectacular pieces of observational evidence supporting the hot big bang model comes from the detection of this cosmic background radiation. Based on the above description, the model predicts two very distinctive properties of this radiation: (1) the radiation should be extremely isotropic (the same from every direction) since it filled the entire universe uniformly, and (2) it should be characteristic of a state of thermal equilibrium, since this is how it formed. Namely, once a temperature is determined, one can predict precisely the intensity of the radiation at every wavelength. Furthermore, since the universe has continuously expanded and cooled ever since the primeval fireball, it is expected that the temperature of this *fossil remnant* radiation today would be extremely low, only about 3 Kelvin. Temperatures in Kelvin are measured from *absolute zero,* which is the lowest temperature possible (about 273 below zero Celsius or 459 below zero Fahrenheit). A degree in Kelvin is equal to a degree in Celsius or to 1.8 degrees in Fahrenheit. This means that the radiation should peak at microwave (radio) wavelengths—it is the kind of radiation that heats food in microwave ovens. The first suggestion for the existence of such a background radiation was made by the brilliant theoretical physicist George Gamow in 1948, and explicit predictions (and even preliminary plans to detect the radiation) were made in the early 1950s by R. A. Alpher, R. C. Herman, and J. W. Follin.

In the early 1960s, a group of researchers at Princeton, led by the outstanding cosmologist Robert Dicke, were in the process of building a microwave antenna to search for the cosmic background radiation, when they were accidentally scooped. Arno Penzias and Robert Wilson of Bell Laboratories in New Jersey were calibrating in 1964 a microwave antenna designed to improve satellite communications. Much to their vexation, they kept finding in their data some background noise, a bit like the hiss one hears on certain radio stations. In spite of all their efforts—which, by the way, included scraping off pigeon droppings from the antenna—this microwave noise just would not go away. Furthermore, the noise remained the same irrespective of when and in which direction they pointed their antenna. It was only through an accidental, but fateful, airline trip that Penzias suddenly realized what he and Wilson discovered. On this trip, Penzias learned from an astronomer who sat next to him of the Princeton's group calculations and experiment. After meetings with the Princeton group, Penzias and Wilson realized that the disturbing noise they found was nothing less than the afterglow of the big bang itself. They received the 1978 Nobel Prize in Physics for this truly momentous discovery.

In 1992, measurements made by astronomer John Mather and his team with the *Cosmic Background Explorer* (COBE) satellite demonstrated beyond doubt two dramatic facts about the microwave background: (1) the radiation is extremely isotropic, the variations from one direction to another amounting only to one part in 100,000; and (2) the radiation has a precise thermal equilibrium spectrum (with an accuracy better than one part in 10,000), corresponding to a temperature of 2.728 Kelvin. These results are in remarkable agreement with the predictions of the big bang model and perhaps more than any other single observation provide strong evidence that the big bang really happened.

In very broad terms, the subsequent evolution of the universe can be described as follows. At the decoupling epoch, once neutral atoms formed, these atoms could collect into gas clouds. The gas clouds later collapsed due to the force of gravity to form stars, galaxies, and clusters of galaxies. It should be noted that at the present time, the exact process of galaxy formation is not fully understood. Furthermore, it turns out that the gravitational influence of matter we cannot see, *dark matter,* plays a dominant role in the formation of structure in the universe (see chapter 4). However, there is no reason to suspect that galaxy and structure formation cannot be accommodated comfortably within the framework of the big bang model.

The behavior of nuclei, atoms, stars, and galaxies is governed by the properties of the basic forces of nature. These, in turn, provide another wonderful example of reductionism and symmetry in action.

Symmetry Breaking: Is the Bread Plate on the Right or on the Left?

At present we recognize in nature four basic interactions, or forces. These are gravitational interaction, electromagnetic interaction, strong nuclear interaction, and weak nuclear interaction. The gravitational force is what holds us on the surface of the earth and what makes apples fall on the heads of people called Newton. It also holds the earth in its orbit around the sun, the gas of the sun itself together, and the sun in its orbit around the center of the Galaxy. The same gravitational force holds the matter of the

Galaxy itself together and many galaxies together in clusters of galaxies.

The electromagnetic force is what holds solids and liquids together and gives them their particular properties. It is responsible for all the chemical processes, for all the phenomena that we call electric or magnetic, and for all the properties of electromagnetic radiation (light, X rays, radio waves, etc.). It was only in the nineteenth century that the physicists Michael Faraday and James Clerk Maxwell recognized that electric and magnetic forces are manifestations of one unified interaction—electromagnetism. Moving electric charges create a magnetic force, while moving magnets create an electric force. In short, when we make our morning toast, the electromagnetic interaction is responsible for holding the toaster together and the toast intact, for the heating of the toaster by electricity, for the fact that its internal wires glow in red light, for the browning of the toast, and eventually for all the processes that occur inside our mouth and stomach when we digest the toast.

The strong nuclear force acts only at very short distances, of the order of the size of the atomic nucleus. It is responsible for holding the protons and neutrons together in the nuclei of atoms. It should be realized that protons have a positive electric charge while neutrons are neutral. Similar electric charges repel each other, and therefore in the absence of an attractive force, the nucleus would immediately disintegrate, with all its constituents flying apart. It is only because the attractive strong nuclear force is stronger than the electromagnetic repulsion that the protons and neutrons in the nucleus cling tightly together.

The weak nuclear force is also of short range, and it plays a role in some radioactive decays known as beta decays (processes in which neutrons are transformed into protons and vice versa). The weak force is also responsible for interactions among elementary particles (subatomic particles that are the basic constituents of matter) that involve a particle called a *neutrino.* This is an elementary particle of extremely low mass, which interacts very weakly with ordinary matter. One of the places where it is produced copiously by nuclear reactions is the center of our sun. The weak nuclear force is weaker than both the strong nuclear and electromagnetic forces. While we see no direct evidence for the weak nuclear force in everyday life, its role in the energy-producing reactions in stellar cores is absolutely crucial.

What can be a better demonstration of reductionism than the long list of phenomena for which the electromagnetic and gravitational forces are responsible? After all, the gravitational force moves things ranging from apples on earth to galaxies in clusters, and the electromagnetic force on one hand prevents us from being able to walk through closed doors and on the other makes computers work. However, physicists are even more ambitious.

Adopting the principles of reductionism and symmetry to their extremes means that in the ultimate theory, physicists would like there to be only *one* basic interaction instead of four. Namely, in such a theory the four forces would be interpreted merely as different manifestations or components of a single force. This desire drives the searches for a *unified theory,* a theory in which all of these forces become one.

What is the meaning of this unification? One may ask, if all the forces are really one, then why do we observe them as four separate forces, with rather different strengths? (For example, the electromagnetic force is about 10^{36} times stronger than the gravitational force.) In order to understand how one basic interaction may manifest itself as four different forces, let us examine what happens to a liquid when it is being cooled down. Inside a liquid all directions appear identical; that is, irrespective of how the liquid is turned, the arrangement of the molecules will look the same. This means that the liquid is completely symmetric under rotation. For example, light propagates with the same speed in all directions inside a liquid. Now imagine that we cool the liquid down to the point where it freezes into a crystal; in the professional jargon, a *phase transition.* At this point, the solid that is formed can assume the configuration of a lattice—like salt crystals, for example—with atoms at the corners of cubes. Inside this lattice, all directions are no longer equivalent. Instead, we would typically have the atoms arranged along three different crystallographic axes of the lattice. Light could definitely propagate at different speeds along the directions marked by the three different axes. Imagine that we were creatures living in a *space* that had the properties of this liquid-turned-solid medium. Then clearly our laws of physics would read quite differently in the liquid phase than they would in the solid phase. In the liquid, we would have a situation in which the speed of light would be given only by one number, while in the solid phase we would have three numbers for the same property. In other words, in the *cold* solid medium we would have three different values describing the physics of light propagation, while we would have only one value in the *hot* liquid medium. This example demonstrates how some quantities that are unified under some conditions can spontaneously separate when a certain symmetry breaking (in this case induced by cooling) occurs. In an analogous way, the fact that we presently observe four different forces, with four different strengths (instead of one), may simply reflect the fact that we live in a cold universe, one in which symmetry breaking has already occurred.

The phenomenon of symmetry breaking is encountered in situations around a round dinner table, in which each guest has a bread plate next to her or his dinner plate. The configuration is entirely symmetric in the sense that all the places around the table are identical and left and right are indistinguishable. Now, as bread is served, imagine that the first diner to reach for the bread uses the bread plate to the left of her or his main plate. This will immediately cause the symmetry to be broken; namely, left and right become distinguishable. In fact, as we know only too well, sometimes two different diners make the choice at the same time, with one choosing the plate on the left and the other the one on the right, leading to a situation in which one diner is left without a plate (while a plate remains unclaimed in another part of the table).

The search for a unification of the forces is not new. Einstein himself devoted a large part of the last thirty years of his life to an attempt to unify the theory of the electromagnetic interaction with general relativity, the theory of gravitation. Namely, he wanted to show that these two forces are merely different manifestations of the same force. He chose these particular two interactions (ignoring the strong and weak nuclear forces) because, in fact, they were the only two known in Einstein's youth. With hindsight we can understand today why he failed in this unification attempt. As I shall describe below, unifying gravity with the other forces is the most difficult part, one that still eludes solution.

The last three decades, however, have witnessed remarkable progress in the hunt for unification of the forces. In particular, in the late 1960s the physicists Steven Weinberg, Abdus Salam, and Sheldon Glashow have shown that the nuclear weak force and the electromagnetic force merely represent different aspects of one force, now known as the *electroweak* force. What has been demonstrated is that the electromagnetic and weak forces merge into one at high energies and, as in the example with the liquid above, they acquired their different manifestations (the symmetry was broken) only after the universe cooled below a certain critical temperature of about 10^{15} Kelvin. That temperature was obtained when the universe was about 10^{-11} seconds old. At that point, particles had kinetic energies of about 100 times the rest-mass energy of the proton. The rest-mass energy of a particle is the energy obtained if we convert the mass of that particle to energy, using Einstein's famous equivalence between mass and energy (which states that the energy is equal to the product of the mass and the square of the speed of light). Since such energies have already been obtained in large particle accelerators, most aspects of the electroweak theory have already been confirmed experimentally. In particular, the theory linked the carrier of electromagnetism, the photon, with three previously undiscovered particles, the carriers of the weak force. The existence of these predicted particles, the W^+, the W^-, and the Z°, was spectacularly demonstrated in 1983.

An even more important prediction is related to the *mechanism* that is responsible for symmetry breaking. The currently favored mechanism is known as the *Higgs field,* after Peter Higgs of the University of Edinburgh, who developed this idea in 1964. The Higgs field acts as a "party pooper" in that when it achieves its lowest energy state (which all systems like to achieve), it breaks the symmetry. The prediction is that Higgs fields are carried by a massive particle, known as the Higgs boson. Experiments with large particle accelerators at the European Organization for Nuclear Research (CERN) in Geneva and at Fermilab in Batavia, Illinois, will, it is hoped, reveal in the near future if the Higgs boson really exists. The goal is to collide particles at sufficiently high energies to create the Higgs boson.

Thus, the general idea behind unification is very simple. All the basic interactions were unified in earlier stages of the universe, when it was much hotter. The universe was highly symmetric, in the sense that interchanging any of the forces among themselves would have resulted in no change. But as the universe expanded and cooled down, at certain critical temperatures (the phase transition temperatures) symmetry breakings occurred, and eventually the four interactions gained their distinctive identities. The universe therefore behaves very much like the liquid I described earlier, and the reason that we observe four different interactions is simply because we live at a stage in which the universe is already very cold. The unification of the strong nuclear force with the electroweak force occurs, according to some theories (known as grand unified theories, or GUTs), only at energies about 10^{15} times the rest-mass energy of the proton. Particles in the universe had such energies only when the universe was less than 10^{-35} seconds old (the *GUTs era*) and its temperature was above 10^{28} Kelvin. Since such high energies cannot be achieved in any accelerator (in fact they are about a trillion times higher than those reached in present-day accelerators), our only hope to test GUTs is either through some predictions they make for our present meager-energy universe or through some relics from the GUTs era. We will see examples for both of these types of tests later.

The above discussion demonstrates something very interesting from a history of science point of view. Astronomy is the oldest of the scientific disciplines (not the oldest profession). It provided the first clues about the existence of laws in the universe. Physics was born when the notion of such laws was pulled down from the heavens to the earth. Now physics pays back this debt, as theories of the fundamental forces apply directly to cosmology and are tested by astronomical observations.

Whether GUTs work or not, the greatest challenge to unification theories is still presented by the goal of unifying gravity with the other interactions. The reasons for the difficulty can be appreciated when one realizes that this unification requires no less than putting together the two most dramatic revolutions in scientific thinking *ever,* under one roof. These two revolutions are quantum mechanics and general relativity. Einstein's general relativity is a theory of gravitation that, for the first time, related the concepts of space and time with those of matter and motion. I will return to general relativity in chapter 5. At this point, let me just note that the theory interprets the force of gravity as a geometrical property, the curvature of a four-dimensional space-time (our three dimensions of space plus a time dimension). The *simplicity* of the central idea of general relativity is so overwhelming that some physicists still consider it to be the most beautiful theory ever formulated. In the words of Hermann Weyl, an outstanding German mathematician and physicist, "a wall which separated us from the truth has collapsed."

Quantum mechanics is the theory that describes the physics of particles and fields (like the electromagnetic field) at the atomic and subatomic levels. It changed the entire vocabulary of physics from one of fully deterministic positions and velocities (or momenta) of particles, to one in which only probabilities for positions and momenta can be determined.

One might have thought that there is no urgency to unite these two theories, to develop a theory of quantum gravity, because their ranges of applicability are generally so different. Quantum effects are completely negligible on the scales of stars and galaxies, where gravity is important. Similarly, the gravitational forces exerted by atoms and subatomic particles (where quantum mechanical effects are important) are minuscule compared to the electromagnetic and nuclear forces. All of this changes, however, in the very early universe, when the entire matter and energy content of our presently observable universe occupied a size smaller than 10^{-33} centimeters. At that time, called the *Planck era* (after the German physicist Max Planck who, in 1900, did pioneering work on the nature of heat radiation), which happened when the universe was younger than about 10^{-43} seconds, one could no longer ignore the fact that gravity needs to be described by quantum mechanics. This description is absolutely essential for unifying gravity with the other forces. A major stumbling block for unification is the fact that no fully convincing theory of quantum gravity exists yet. Nevertheless, there does exist already a basic direction that appears very promising in its potential ability to unify all the forces. Theories that use this basic approach are known as *string theories.* In these theories, the basic physical entity is not a particle, like an electron, but rather a quantum-mechanical *string.* We can imagine these strings to be tiny (almost 10^{20} times smaller than the atomic nucleus), one-dimensional rips in the otherwise smooth fabric of space. In this sense, string theories are analogous to general relativity, in that instead of talking about forces they talk about properties of space itself. The strings can be closed, like an infinitely thin rubber band, or open, with both their ends free. Because they are so tiny, loops of strings appear like pointlike particles, when not probed too closely. Like a guitar string, the strings can vibrate in a variety of "tones," and various particles correspond to different tones. Thus, the vibrations of looplike strings produce the mass and the charge of particles, rather than music. For example, calculations by John Schwarz of Caltech and Joel Scherk of the École Normale Supérieure in Paris have shown that one of the tones of a closed string would appear like a particle with zero mass, which has a spin twice that of the photon (twice the angular momentum), the carrier of electromagnetism. The remarkable thing is that these are exactly the expected properties of the *graviton,* the carrier of the gravitational force. Since the existence of the graviton seems to be an *unavoidable* feature of all string theories, one is inclined to say that string theories predict gravitation to exist.

At present, in spite of some impressive successes, string theories still have a long way to go before they can be considered an ultimate unified theory of all the forces. Nevertheless, the progress is continuous, with one of the most recent developments being the realization that a number of string theories are actually a part of a unified framework, presently dubbed somewhat mysteriously M-theory (it is not exactly clear what the "M" stands for; string theorists suggest "mystery" and "mother" as two of the possibilities). One of the reasons for the relatively slow progress is the fact that the mathematical tools needed for the solutions are so advanced that they have to be invented along the way.

An intriguing possibility that string theories raise is that the subatomic elementary particles are in fact *extremal black holes.* Black holes in general are objects that produce such a strong gravity that matter or energy cannot escape from them (I will return to this topic in chapter 8). Extremal black holes are tiny black holes, the size of subatomic particles, which have an electric charge and the minimal possible mass. Significant progress toward a correspondence between strings and black holes was made in 1995 by Andrew Strominger of the University of California at Santa Barbara, and a direct connection between black holes and elementary particles has been established by Strominger, Brian Greene of Cornell University, and David Morrison of Duke University. This fascinating development suggests that even such seemingly different entities as black holes and subatomic particles may ultimately be simply different manifestations of the same basic entity, namely, strings. It is beyond the scope of this book to discuss this topic further; I refer the interested reader to Brian Greene's excellent book *The Elegant Universe.* Interestingly, in an interview with science writer Timothy Ferris (author of *The Whole Shebang*) in 1983, Strominger said: "I don't believe anyone finds out anything about the universe except how beautiful it is, which we know already."

From the point of view of the present book, it is not important if string theories or the M-theory (or indeed other theories such as one dubbed "loop quantum gravity") will provide all the answers. The important point is that many if not most physicists *have no doubt* that such a unified theory exists, and that when it is found, its central idea will undoubtedly have simplicity and symmetry written all over it! This explains why some of the best scientific minds of today, and in particular Edward Witten, an outstanding mathematical physicist at Princeton, devote all their energy to the search for such a theory. A few years ago I invited Witten to give a talk at the Space Telescope Science Institute. As we discussed some of the recent astronomical findings, I asked him whether he was not worried by the fact that string theories may not be able to produce directly testable predictions. His reply was immediate: "We predict gravity!"

What would have happened if symmetry breaking had not occurred in the universe and the interactions had stayed unified forever? There would have been only one force and one type of elementary particle. In such a universe the hydrogen atom (or any other atom) would not have existed. There would have been no galaxies, no flowers, no us. Does that mean that our universe, in which symmetry breaking has occurred, is not beautiful? Not at all, unless you identify boredom with beauty. We have to remember that the beauty lies in the fact that a *wealth* of phenomena emerged from a single primeval force, rather than in the mere existence of such a force. Similarly, the Russian playwright Anton Chekhov sometimes used boredom as a dramatic element to describe a complex human condition, but never as the only constituent of a play.

Conjugations in the Mirror

One of the major successes of the incorporation of grand unified theories within the big bang model has been the understanding of the fact that our universe appears to contain *matter* and not *antimatter.*

From experiments with large particle accelerators and from the theory of elementary particles we know that every elementary particle has a corresponding antiparticle. The antiparticle has the same mass as the particle, but the opposite electric charge (of the same magnitude). Thus, the antiparticle of the electron, called a positron, has the same mass as the electron, but is positively charged. Indeed, the discovery of the positron, in 1932, was one of the first successes of quantum electrodynamics—the theory describing the behavior of electrons and light. Similarly, there exists an antiproton that is equal in mass to the proton but is negatively charged (in fact, the proton consists of three quarks and the antiproton of three antiquarks). All of these antiparticles can be produced in pairs, with their corresponding particles, in high-energy experiments—for example, in which protons are smashed into other protons with energies exceeding the rest-mass energy of the pair that needs to be produced. It is also known that in the same way that the hydrogen atom consists of a nucleus (which is a proton) and an electron orbiting the nucleus, one could construct an antihydrogen atom, which would have an antiproton at the center, with a positron orbiting it. In this way, in fact, an antiatom to any atom could exist. Here, however, comes a truly surprising observational fact. As far as we can tell, essentially *all of the matter in the universe is ordinary matter;* namely, it consists of protons and electrons rather than antimatter. Antimatter constitutes less than a millionth (and perhaps even much, much less) of the total amount of matter.

This conclusion rests on the known consequences of collisions between matter and antimatter. When protons and antiprotons collide at not too high energies, they completely annihilate each other, with the energy being emitted as a burst of pure radiation (in the form of gamma rays). Consequently, collisions between matter galaxies and antimatter galaxies would have resulted in total annihilation, accompanied by brilliant bursts of gamma rays. No such phenomenon has ever been observed.

The overwhelming prevalence of matter over antimatter in our universe seems, at first glance, to be in serious conflict with the unifying, reductionist principle of the forces. If the early universe had been precisely symmetric, it would be expected to contain equal numbers of particles and antiparticles. However, if this were the case, we would not have been here to ponder over it. A universe with equal numbers of particles and antiparticles would have ended up, following all the annihilations, filled with radiation, which would have subsequently cooled down as the universe expanded, but with no atoms.

Hence, such a universe would have contained no planets, no stars, and no galaxies. How can we then explain the fact that our present observable universe contains some 10^{78} protons, but essentially no antiprotons? The answer lies in a tiny violation of the perfect symmetry between particles and antiparticles called *CP* violation.

In order to explain this concept I will make use of Figure 6a. Suppose we apply to this figure the following operation: *Interchange black areas with white areas.* This will result in Figure 6b, which is clearly different from Figure 6a (e.g., the black is now on the right instead of on the left). We can therefore say that Figure 6a is *not* symmetric under the above operation. Let us now perform another operation to Figure 6a: namely, *reflect the image (as in a mir-*

Figure 6a

Figure 6b

ror). This will again result in Figure 6b, and so we find that the original image is also *not* symmetric under the second operation.

Imagine now however that we perform the two operations consecutively; that is, in Figure 6a we first interchange black with white (I'll denote this operation by *C*) and we then reflect the image (basically reverse left and right; I'll denote the second operation by *P*). Clearly, in this case the figure will remain *unchanged.* Thus, while the figure is symmetric under neither *C* nor *P*, it *is* symmetric under the combined operation *CP.*

Incidentally, a number of the drawings by the fantastic Dutch graphic artist M. C. Escher possess the above symmetry. For example, he has one drawing in which black riders are moving to the left, with all the spaces among them filled precisely with identical white riders moving to the right. That drawing is not symmetric under *C* or *P* (as defined above), but is symmetric under *CP.*

Reactions among elementary particles obey similar symmetries. For example, the operation known as Charge Conjugation (and denoted by *C*) simply reverses the charges of the particles. It is known experimentally that the nuclear weak interactions among particles do not remain exactly the same when the charges of the particles are reversed; namely, *C* is violated. Similarly, the parity (denoted by

P) operation simply reverses left and right (like a mirror reflection). If the interactions were symmetrical under this operation, then mirror reflections of particle interactions would have been identical. Again, however, it was found experimentally that mirror reflections of elementary particle weak nuclear interactions are not the same. Thus *P* is violated, too. For some time, particle physicists still thought that as in the above example, the interactions are symmetric at least under the combined operation *CP* (reversing the charges and left and right). However, in the mid 1960s, an experiment by particle physicists Jim Cronin, now at the University of Chicago, and Val Fitch of Princeton University, which studied the decay of particles known as K-mesons (or kaons), revealed to everybody's astonishment that in about 0.2 percent of the cases the decays violated the *CP* symmetry. That is, unlike in the above example, the decays were not the same even if one *both* changed the charges and reversed left and right. Since in the early universe the weak and strong interactions were unified, the immediate implication of this discovery is that reactions in the GUTs era and immediately following it, when the universe was younger than about 10^{-32} seconds, violated *CP*. This allowed for the possibility of creating a slight excess of particles over antiparticles, an idea first presented by the Soviet physicist Andrei Sakharov. One may ask how slight that excess had to be to account for the fact that today we observe only protons and neutrons and no antiprotons and antineutrons. Amazingly enough, it was sufficient that roughly for every *3 billion* antiparticles created, there were *3 billion and one* particles!

To demonstrate that this tiny excess is all that was needed, we have to examine the consequences of such an excess. As the universe expanded and cooled down, all the antiparticles were annihilated with the corresponding particles, generating photons of radiation. But for every 3 billion antiparticle-particle pairs that disappeared, one unpaired particle survived. Since every annihilation creates two photons, we would expect to find in today's observable universe more than 1 billion photons for every proton. The number of photons in the 2.7 Kelvin cosmic background radiation is indeed more than a billion times larger than the number of protons in the observable universe. We therefore come to the realization that *the entire materiate world, from galaxies to humans, owes its existence to that tiny excess of particles over antiparticles,* which in itself was a consequence of a tiny violation of perfect symmetry.

The fact that some small violation of precise symmetry can actually enhance the beauty is not something that is special to physics, as anyone looking at a rose has immediately realized. Many works of art use slight departures from symmetry in constructing what is known as the "visual form," namely, those features that give the work its organization and coherence.

A magnificent example is provided by a painting called *Truth* (now in the Kunsthaus, in Zurich), painted in 1903 by the Swiss symbolist painter Ferdinand Hodler. In that painting, truth is personified by a beautiful nude woman in the center, facing the viewer, with her arms extended as if to embrace all her surroundings. On each side of the woman there are three black-draped phantoms of deceit and evil, facing away from the viewer. The two groups of demons are very similar, but not identical, which literally forces the viewer to examine each one of them in detail. It is interesting to note that Hodler used similar constructions, and a nude woman as signifying spiritual purity, in his paintings *Day* and *Art,* implying perhaps that he regarded those concepts as components of truth. The physicist Philip Morrison from MIT says in his book *Nothing Is Too Wonderful to Be True,* "What we regard as highly satisfying works of art, even many natural things of beauty, contain broken symmetries."

The fact that at energies higher than those prevailing when the universe was about 10^{-32} seconds old an excess of particles over antiparticles could be generated immediately suggests that in the basic interaction, the difference between particles and antiparticles is not a strictly conserved number. The baryon number of a particle is defined as the number of quarks it contains, minus the number of antiquarks, all divided by three. For example, a proton contains three quarks and no antiquarks, so its baryon number is one unit. An antiproton contains three antiquarks (and no quarks), so its baryon number is negative one unit. In all the experiments and processes that have been studied to date, the baryon number was found to be conserved. In addition, the proton happens to be the lightest baryon that exists. Consequently, if baryon number conservation were to hold absolutely, then the proton would never decay, since particles can decay only to lighter particles, but this would change the total baryon number. However, at GUT energies, the baryon number is not exactly conserved. In fact, at temperatures exceeding 10^{27} Kelvin, processes that change the baryon number are expected to be quite common. A dramatic prediction of baryon nonconservation is that not only are diamonds not forever, even the proton, the most fundamental building block of ordinary matter, is unstable and eventually decays. There is no immediate cause for alarm, however. The presently determined limits on the proton's lifetime put it at longer than 10^{32} years. This means that even if we lived for 2,500 years, only one proton in the atoms in our body would have decayed. This finite lifetime does offer, however, the possibility of an experimental test, since it means that in about one hundred tons of hydrogen one atom will decay every year. A few experiments designed to detect such decays (if they occur) have been operating in the past 20 years. These experiments have already ruled out the simplest version of GUT. According to this minimal version, the proton should have decayed to a positron and a particle called pi zero in less than 10^{30} years. However, three experiments have been operating long enough to demonstrate that the proton does not decay to these particles even in 10^{33} years. The precise implications of these results for GUTs in general are not known yet.

This reminds me of a cartoon I once saw in which a scientist leaving his lab is seen saying: "First I hear that marriage is unstable, then that the economy is unstable, and now that even the proton is unstable, so what is the wonder that I am depressed?"

As I described above, *CP* violation was discovered in kaon decay; but kaons, which are relatively light particles, are too simple to flesh out the entire picture concerning *CP* violation. The hope is that the decays of heavier particles, known as B-mesons, will provide a far richer menagerie of particles and thereby a more complete picture of the source of *CP* violation. Two large particle accelerators, known as B factories, started producing in the spring of 1999 large numbers of B-mesons and to study their decays. One of these experiments is at the Stanford Linear Accelerator Center (SLAC), and the other is at the High-Energy Accelerator Research Organization (KEK) in Tsukuba, Japan. In both machines electrons and positrons are whirling around in opposite directions at nearly the speed of light. When bunches of these electrons are made to collide with the positrons, pairs of B and anti-B-mesons are produced copiously. The hope is that besides possibly unveiling the mystery of *CP* violation, these new experiments will also provide new insights into the nature of the standard model of elementary particles that unifies the forces. Incidentally, the large detector at SLAC, which will record the paths of the particles being produced and determine their energies, is called Babar, after the elephant in Laurent de Brunhoff's books.

Backward, Turn Backward, O Time in Your Flight

The predominance of matter over antimatter in the universe probably originates from the fact that the basic forces in nature violate slightly the *CP* symmetry. This violation, however, has another intriguing consequence. It implies that the reactions among particles would not remain the same if time were to flow backward—they distinguish between past and future! This is our first encounter with the existence of an "arrow of time," a property of the fundamental laws that determines a direction for the flow of time. This may appear somewhat puzzling, since, as I noted before, the basic *equations* that describe the laws of physics are symmetric with respect to time—namely, they remain unchanged if we reverse the flow of time. Before I attempt to explain this apparent paradox, however, let me answer right away an even more burning question. Could it be that this arrow of time is responsible for the fact that we humans have such a strong sensation of the flow of time? And such an overwhelming distinction between past and future? The answer is: almost certainly not. Remember that this asymmetry in the flow of time is a consequence of *CP* violation, which occurs only in some very small fraction of the cases of the decay of rare particles, the K- and B-mesons. It is difficult to see how this small effect could generate something as definitive as our subjective arrow of time. Thus, we will probably have to seek elsewhere for the origin of our subjective sense of the direction of time. Now, let me go back to the question of how it is possible for the reactions to distinguish between past and future while the equations describing the theory are symmetric in time. This is again related to symmetry breaking. The process may be best described by the behavior (in terms of direction in space) of an ordinary bar magnet. Inside a magnet, there are many microscopic atomic magnets. The equations describing every such tiny magnet and its interactions with all the other tiny magnets are *completely symmetrical* with respect to every direction in space. That is, the equations do not distinguish among the different directions. Yet any *individual solution* to the equations, describing the direction of an individual tiny magnet, *violates the symmetry,* because it points in a *definite* direction. The *entire collection* of all the solutions is *symmetrical* again, because there is a tiny magnet pointing in *any* direction. These circumstances result in a situation in which the slightest external influence that favors a given direction will result in the magnet pointing in that direction. The situation is somewhat similar to a cocktail party in honor of some distinguished guest. As long as the guest of honor is not there, every person is facing some definite direction, but the collection of all invitees does not favor any direction in particular. As soon as the guest arrives, however, everybody turns to face the same direction. In the case of the arrow of time, the equations are symmetric with respect to time, yet they probably have two solutions, each one with the opposite direction of time. Therefore, the collection of *all* the solutions probably does not have a preferred direction of time. However, due to symmetry breaking, only one of these solutions apparently materialized in nature, thus giving time a *specific* direction.

Probably the best-known arrow of time, other than our own subjective arrow that distinguishes between past and future, is the thermodynamic arrow. In thermodynamics (the science of heat phenomena) there exists the concept of *entropy,* which is a measure of the *disorder* of any physical system, be it at the molecular or the macroscopic level. For example, if an amount of gas is all squeezed at one corner of the room, its entropy is lower than if it is dispersed throughout the entire room, since the system is more ordered in the first case.

Similarly, if carbon, hydrogen, oxygen, iron, and other molecules are wandering aimlessly through interstellar space, their entropy (disorder) is higher than if all of these molecules combine in a perfect way to make a human body. In everyday life, if a collection of books is scattered all over the floor, its entropy is higher than if the books are all arranged on shelves, in alphabetical order by author.

In 1865 the German physicist Rudolf Clausius formulated the *second law of thermodynamics,* which states that the entropy of any isolated system can never decrease. Namely, systems can only become more disordered with the passing of time. Anyone who ever watched a teenager's room over a period of time (or my office, for that matter) will agree that the second law of thermodynamics works. Similarly, if a gas is confined by a hermetically closed partition in one corner of the room and then the partition is lifted, the gas will tend to flow until it eventually fills the entire room homogeneously, and has a uniform temperature. This is a state of *thermodynamic equilibrium,* and in this state the entropy of the system is maximal. The second law is also responsible for the fact that a hotter body and a colder body that are in contact eventually equilibrate their temperatures by heat flowing from high to low temperature.

The second law of thermodynamics defines a clear direction of time for isolated systems, set by *increasing disorder.* In systems that are not isolated—namely, that are in contact with other systems—order can actually emerge out of disorder. For example, if you watch from above oil heating in a round open pot, you will see that the homogeneous and amorphous oil suddenly takes the form of hexagonal connected cells, a phenomenon first discovered in 1900 by the French physicist Henri Bénard. Order in this case results from the external agent of heating.

There are two other arrows of time that I would like to mention, since they characterize the universe at large. One is related directly to the expansion of the universe. The expansion itself defines a cosmological direction of time, since if we compare the size of the universe at two different times, as with an inflating balloon, we can tell which came first. A second arrow is related to the appearance of order in the universe. At the big bang the universe was dense, hot, and essentially featureless. But out of this chaos, order managed to crystallize somehow, first in the form of atoms, then molecules, and eventually stars, galaxies, and clusters of galaxies. Thus, there appears to be an arrow of increasing universal order. This raises two immediate questions: First, isn't this appearance of order in the universe in contradiction to the second law of thermodynamics, which envisages an ever-increasing messiness? Second, are all of these arrows of time—the subjective, the thermodynamic, the cosmological, and that of universal order—somehow interlinked? Or is it a mere coincidence that they all appear to point in the same direction?

Before I attempt to answer these questions I would like to describe a wonderful thought experiment designed by the nineteenth-century British physicist James Clerk Maxwell (who also formulated the laws of electromagnetism). It involves a hypothetical demon that spends its time sorting. Maxwell's experiment consists of a vertical, cylindrical container, which is full of gas and is separated by a partition into two identical halves (an upper and a lower half). In the partition, there is a small door, which is operated by the little demon. At the beginning of the experiment, the temperature of the gas in the two halves is precisely equal, which means that the *average* speed of the gas molecules is the same. The demon is able to determine the speeds of individual molecules, and by opening at will the partition door, it allows faster-than-average molecules to pass from the lower half to the upper half and slower molecules to pass downward. Thus, the demon manages to collect faster molecules (which correspond to a higher temperature) in the upper half and slower ones (lower temperature) in the lower half. In other words, due to the demon's activity, the upper half continuously warms up while the lower one cools down. This, however, *lowers* the entropy of the system—the system becomes more ordered (and heat flows from the cold gas to the hot gas), in apparent conflict with the second law of thermodynamics. However, in calculating the total change in entropy we must include the demon itself. Since we are quite convinced that the second law works (and the total entropy increases), the increase in the entropy caused by the demon's actions must somehow be at least as large as the decrease in the entropy of the gas. The solution to this problem was suggested by works of the mathematician Claude Shannon and the physicist Léon Brillouin. First, they showed that having *information* on the system (and recording it) at the molecular level can reduce the entropy (the disorder) in spite of the second law. Microscopic information is thus equivalent to order. Brillouin further demonstrated that in the process of gathering information, which the demon needed in order to know which molecules to allow where, more entropy was generated than the entropy reduced from the gas by ordering it. Thus, the total entropy of the universe *increased* in agreement with the second law of thermodynamics, even though the entropy of the particular system decreased. The second law of thermodynamics therefore states that the entropy of the universe as a whole can never decrease. The universe has to become increasingly disordered.

It is relatively easy to understand how, starting from an *ordered state,* disorder could develop. Humpty Dumpty, who "sat on the wall" in Alice's adventures in *Through the Looking Glass,* had all his pieces in their proper place, and there is only one possible arrangement in which all of these pieces fit perfectly together to form the intact Humpty. On the other hand, once he "had a great fall," there were very many arrangements in which his scattered pieces could be lying on the ground. Thus, there is always a high probability of disorder emerging from order, because there are many more possibilities of disorder. Similarly, while there is an infinite number of ways in which a collection of books can be scattered on the floor, there is precisely one way in which they are arranged alphabetically. Here, however, we encounter an apparent paradox: If at the big bang matter was in thermal equilibrium, and disorder (entropy) was *maximal,* how is it possible for the entropy to still grow in the universe?

The answer could be related to the fact that the universe is *expanding.* The maximal value that the entropy can attain depends on the size of the system. In a larger container, the entropy of a gas in thermal equilibrium is larger than in a smaller one. The expansion of the universe thus can allow the entropy to grow. This could explain why the cosmological and thermodynamic arrows point in the same direction—the increase in the size of the universe is accompanied by an increase in the entropy. Furthermore, we can now also understand how it was possible (in spite of the second law) for order to appear in the universe. When a given system is in thermodynamic equilibrium, its entropy, or disorder, is maximal. This means that as long as the system is *not* in thermodynamic equilibrium, the disorder is less than maximal; thus, there is some *order* in the system. In fact, the farther the system is from equilibrium, the more ordered it is. In the example of the gas in a container, the equilibrium corresponds to the gas filling the entire container homogeneously. The more concentrated the gas is in a small part of the container, the more ordered the system is. Imagine now that you start with the gas in the container being in thermal equilibrium (homogeneously spread out), but that the container itself is capable of expanding. If the container expands much faster than the gas can flow, you may even wind up with the gas suddenly occupying a very small volume of the vastly expanded container. Namely, order has appeared.

The situation with the universe could be quite similar. At the beginning, the universe was in thermodynamic equilibrium, and its entropy was maximal, corresponding to total disorder. As the universe expands, two things happen: The matter and radiation continuously become more dilute and cool down, and the maximal value that the entropy can attain is continuously increasing (with the increasing size). Due to the dilution and cooling, the rates of all the atomic and nuclear processes that result from collisions among particles are reduced, as the matter spreads thinner and the random motions of these particles become less and less agitated. From a certain point on, these processes could not keep up with the expansion, and the actual entropy of the universe started falling behind the maximum possible entropy. The increasing gap between the actual disorder and the maximally possible disorder simply means that order could appear, on different scales. Gravity also played a crucial role in the emergence of cosmic structure. Gravity is responsible for the fact that small clumps of matter became increasingly denser. Thus, it was gravity that operated on all the small departures from a uniform density that existed in the very early universe and made them form stars, galaxies, and clusters of galaxies.

The famous British physicist Roger Penrose has a different view on the origin of the thermodynamic arrow. In Penrose's picture, gravity plays a dominant role in determining the direction of the entropy increase. As I noted before, it is easy to understand that entropy, or disorder, should increase if you start from a *highly ordered initial state.* Penrose points out that in the early universe, the distribution of matter was in fact *extremely smooth and uniform,* with no big mass concentrations. Had such concentrations existed very early on, they would have collapsed to form black holes—masses crushed by gravity to extremely compact sizes. Black holes, it turns out, have an enormous entropy, one that far exceeds that of matter spread uniformly. Therefore, Penrose concludes, the entropy of the early universe was in fact *very low,* much lower than its possible maximum value. From this highly ordered and smooth state, the entropy could only increase, with every small clumpiness in the matter distribution being enhanced by gravity. The second law of thermodynamics according to this picture is simply a consequence of the homogeneity of the early cosmos.

Finally, how is all of this related to our subjective arrow? How is it that examining three autoportraits of Rembrandt from different periods in his life, we can immediately tell the order in which they have been executed? Why is it that we fear death, but do not fear the time before we were born? Many researchers tend to associate the subjective arrow of time with memory. Already in *Confessions* (written in 397), Saint Augustine writes: "In thee it is, O my mind, that I measure my times." The seventeenth-century Jewish philosopher Benedict (Baruch) de Spinoza expresses a similar view: "Duration is the attribute under which we conceive the existence of created things." Memory means that *information* needs to be recorded in the brain. As we have learned from the example of Maxwell's demon, information is the equivalent of order, which reduces the entropy of the given system (the particular region of the brain in this case). However, according to the second law of thermodynamics, this means that at the same time entropy must be generated elsewhere in the brain or in the body (e.g., in the form of increased heat), with the net result being that the total entropy of the universe increases. Thus, *if* the subjective arrow of time is indeed associated with the accumulation of memories, then this arrow points in the same direction as the thermodynamic arrow, since the increase in entropy and the gathering of memories go hand in hand.

I do not wish to elaborate any further on this topic here, except for noting that, as the above discussion shows, the increase in the messiness of the world, the appearance of cosmic order, and even our own distinction between past and future may all at some level be related to the expansion of the universe and to the forces that govern its behavior.

Hubble's discovery has literally and figuratively expanded our universe. My friend Brian Warner, an astronomer at the University of Cape Town, wrote a small book of humorous scientific poems called *Dinosaur's End.* In it, he has an amusing poem called "A Shrinking Feeling," which goes like this:

The Universe expanding is common understanding,
But another fact that scientists give
Is that motion's purely relative
Which sets one to thinking:
Could *we* be shrinking?
And if that's so,
How long to go
Until it
Finis
Is
?

The poem could almost be taken as reflecting our emotional reaction to the increasing applicability of the Copernican principle, to the apparent decreasing significance of our physical existence in an expanding universe.

The 1920s witnessed the discovery of the expansion of *space* itself. However, intriguing findings concerning the nature of *matter* in the universe were soon to follow.

4

The Case of the Missing Matter

In Sir Arthur Conan Doyle's short story "The Adventure of the Blue Carbuncle," Dr. Watson comes to visit his friend Sherlock Holmes and finds him examining a hard felt hat. At the conclusion of this examination, Holmes is able, to Watson's complete astonishment, to make the following inferences about the unknown owner of the hat: that the man is highly intellectual; that he was fairly wealthy within the last three years, although he has fallen more recently upon worse days; that he had foresight, but has lost some of it, probably due to drinking; that his wife ceased to love him; that he is middle-aged; that he goes out little; and that it is extremely improbable that he has gas laid in his house. Responding to Watson's amazement, Holmes then starts explaining, one by one, how he reached those conclusions. For example, he deduced the man's intellectual ability from the unusually large size of his head. He concluded that he goes out little from the fact that the dust on the hat was not the gritty, gray dust of the street, but rather the fluffy dust of the house; and so on. Later, when Watson insists, "But his wife—you said that she had ceased to love him," Holmes replies: "This hat has not been brushed for weeks. When I see you, my dear Watson, with a week's accumulation of dust upon your hat, and when your wife allows you to go out in such a state, I shall fear that you also have been unfortunate enough to lose your wife's affection."

In fact, the soundness of most of Holmes's deductions was seriously questioned later by many experts (e.g., a large head does not necessarily imply superior brainpower). However, it is certainly the case that careful deductions can be extremely informative.

Modern "detectives" have a rich arsenal of scientific tools at their disposal. For example, the Rembrandt Research Project is a huge, concentrated effort to establish authenticity for all the works attributed to the Dutch master Rembrandt van Rijn. The fact that Rembrandt had quite a number of talented students and followers, many of whom were apparently quite happy to imitate the master, makes this effort very difficult. The point is that the researchers do not rely on *visual* impressions alone. The paintings are exposed to X rays, infrared photography (which discovers black pigments in covered drawings), ultraviolet microscopy (which uncovers transparent lake pigments), and even thermal neutrons (a technique known as autoradiography). In one of Rembrandt's many self-portraits, one from 1660, these powerful techniques revealed a very soft preparatory sketch and the amazing spontaneity of Rembrandt's brush strokes, both of which produce an image reminiscent of Cézanne.

The lesson here is an important one: We cannot rely on our eyes alone to uncover the true nature and beauty of the universe. Start with a simple example: Anyone who ever went to the seashore on a moonless night had the experience of being able to see only the white foam on the crest of the waves, while the dark, main body of the water remained totally invisible. Imagine we are standing in front of such a sea, and we have never seen a sea before; is there any way for us to deduce more about this phenomenon? A good start would be not to rely on sight alone. For example, we could use the sound of the waves breaking on the shore to infer something about the amount of fluid involved in a wave or about its height. This way we may conclude that there is more to the wave than the foam at its crest.

The universe is similar in some respects to the dark sea. Most of the matter in the universe is dark. The stuff that is responsible for holding galaxies and clusters of galaxies together is matter we cannot see. So how do we know that it really exists? And if it does, what is it?

This cosmic mystery started occupying astronomers in the 1930s and the case is still open. Along the way, a variety of suspects emerged as potential dark matter candidates. Identifying these suspects required quite a bit of detectivelike investigation and deduction in itself. I will now describe briefly the story of three such investigations, because they are intimately related to properties of the universe that are relevant to the discussion of beauty.

Mind over Matter

The first case of a brilliant deduction has to do with radioactive beta decay. Beta decay is a process in which the nucleus of an atom alters its composition to achieve a greater stability. Namely, the nucleus decays from an unstable state to a more stable state. In the process, the nucleus emits an electron. This in itself was considered puzzling at first, since there were strong arguments suggesting that electrons cannot exist inside the nucleus. This difficulty has been disposed of, however, by realizing that beta decay is in fact the spontaneous conversion of a neutron in the nucleus into a proton and an electron, with the electron leaving the nucleus immediately after being created. However, a much more serious difficulty has surfaced. Beta decay apparently *violates no less than the conservation laws of linear momentum, angular momentum, and energy.* For example, in some beta decays the directions of the emitted electron and of the recoiling nucleus could be observed. Now, since the decaying nucleus was originally at rest (*zero* linear momentum), conservation of linear momentum required that the electron and the recoiling nucleus would fly in *precisely opposite directions,* so that their momenta would cancel out. This, however, was almost never observed. A suspicion that the missing energy was lost during collisions between the emitted electron and the atomic electrons surrounding the nucleus was also refuted experimentally in 1927. As I have noted before, conservation laws, which are the result of symmetries, the main ingredient of beauty, are too dear to the heart of every physicist for their violation to be taken lightly. To save the laws, the theoretical physicist Wolfgang Pauli proposed in 1930 that there is another particle emitted in beta decays, which eludes detection. This elusive particle had to have a very small or zero mass and no electric charge, in order to be able to escape being detected. The particle, later christened *neutrino* ("little neutral" in Italian) by the physicist Enrico Fermi, was supposed to carry off the excess energy, linear momentum, and angular momentum that were needed for these quantities to be conserved. The initial experiments aimed at detecting the neutrino, first indirectly, were carried out from 1936 to 1939, and they gave only weak evidence. Convincing, but still indirect, results were obtained in 1942. Direct proof for the existence of the neutrino had to wait till 1956, when Fred Reines and Clyde Cowan detected the flux of neutrinos produced near a uranium reactor. In fact, at present we know that neutrinos exist in three varieties, corresponding to three generations of particles. One is associated with the electron, one with a particle called a muon, and one with a particle called a tauon. The moral here is that when conservation laws are at stake, one should not refrain from bold inferences, such as the existence of unseen particles.

While the above story describes the discovery of a previously unknown particle due to the violation of conservation laws, the next one is related directly to a symmetry.

Particle accelerators are huge devices in which subatomic particles are accelerated to very high energies using electric and magnetic fields. These particles are then directed to collide with other particles, and the detected results of these collisions are used to study the nature of the basic interactions that are involved in the collision. In the late 1950s, experiments with particle accelerators started to reveal the existence of many previously unknown particles. Many of these particles could apparently be divided into families of similar properties. For example, the familiar proton and the neutron seemed to belong to a family of eight particles, all with the same spin angular momentum and with similar masses. Since many of the better-known particles fell into eight-member families, particle physicist Murray Gell-Mann borrowed a term from Buddhism and called this classification scheme "the eightfold way."

In an attempt to put some order into the zoo of new particles, physicists searched for *symmetries* that would permit a clear division into families. What does this symmetry mean? The symmetries that we have already encountered, such as the symmetry under rotation and translation, had to do with changing our point of view in space. The symmetry here has rather to do with the identity of the various particles. One such remarkable symmetry that is seldom mentioned is expressed by the fact that *all* the electrons in the universe are *precisely* identical. Physicists, however, were seeking more encompassing symmetries. Particles would belong to the same family if the laws of nature do not distinguish among them. Furthermore, quantum mechanics allows us to exchange a particle in the family with some arbitrary mixture of other particles in the family. While this may sound bizarre, in quantum mechanics a particle may be in a state in which it is neither precisely one particle nor another, but rather some mixture of both. The mathematical theory that deals with all the possible symmetries and classification schemes is known as *group theory.*

In 1960, Murray Gell-Mann and another particle physicist, Yuval Ne'eman, found (independently) that most of the known particles could be accommodated within the structure of one particular classification scheme. Their theory had one very clear prediction, however. There was one group of nine known particles, which according to the theory *was missing a tenth member.* That missing particle, the existence of which was predicted by the theory, was called *omega minus.* Furthermore, Gell-Mann managed to estimate theoretically what mass that particle should have. The omega minus was eventually discovered in an experiment at Brookhaven National Laboratory on Long Island in 1964, and it had the mass predicted by Gell-Mann!

This second example provides us with different circumstances under which the existence of previously unknown particles can be predicted. Namely, these particles are *required* to exist by some fundamental symmetry property of the basic forces.

The third and last example is much simpler; it deals with matter the existence of which is deduced by its influence.

The planet Uranus was discovered in 1781 by the British astronomer William Herschel. At first he was puzzled by the appearance of this object, described in his words as "a curious either nebulous star or perhaps a comet." However, shortly thereafter, he concluded that the object moved too slowly to be a comet and thereby established its planetary status. Incidentally, he originally wanted to name it Georgium Sidus ("George's Star" in Latin) after King George III (I am not sure I would have been amused by a planet called "George"). Eventually, however, a suggestion by astronomer Johann Bode prevailed, and the planet was named after Saturn's mythological father. Soon after its discovery, astronomers charting the orbit of Uranus found small discrepancies between the predicted and the observed positions. Basically, they were unable to determine a simple elliptical orbit that would fit the observations. The discrepancy became uncomfortably large about half a century later as the planet continued to move.

Since besides the gravitational pull of the sun, which has the main effect in determining the orbit, the gravitational forces of the other planets could also introduce small deviations, astronomers set out to calculate these smaller effects. In September 1845 the British astronomer John Adams managed to prove that the deviations in Uranus's orbit *could not* result merely from the gravitational forces of the other known planets and that there had to be another, previously undetected planet in the solar system. Adams was able to calculate the new planet's mass and its expected position in the sky, but he was unable to convince British astronomers to look for it. In June 1846, the French mathematician Urbain Leverrier, working independently, reached the same results as Adams. This finally provoked an unsuccessful search by British astronomers in the summer of 1846. The new planet, later named Neptune, was discovered in September 1846 by the German astronomer Johann Galle.

This last example shows how one can deduce the presence of unseen matter from orbits, simply by using the known effects of gravity. I will soon show how similar techniques are used to detect vast amounts of dark matter in the universe. Before I do that, however, I would like to share with the reader at least some of the excitement I feel when faced with such theoretical predictions as that for the existence of the omega minus particle. The exhilaration here is double, because this is not merely a case of a theoretical prediction that is confirmed by experiment, of which there are many examples. This is a case in which a remarkable achievement of the human mind has been the direct consequence of being guided entirely by symmetry, *by the intuitive belief in an underlying beauty.* Yuval Ne'eman, one of the two physicists who predicted the existence of the omega minus, was for many years in the Israeli army, before becoming a particle physicist. He once told me half seriously that so strong was his conviction that the omega minus had to be discovered that he considered returning to the army if the experimental search had failed.

The Forces of Darkness

In the 1930s, the Swiss-American astronomer Fritz Zwicky published what can be regarded as groundbreaking results in the study of clusters of galaxies. He suggested that most of the matter in clusters is dark and completely invisible.

Zwicky's logic was really very simple. In clusters of galaxies, he argued, the force of gravity exactly balances the tendency of the galaxies to fly apart due to their own motions. If gravity was any stronger, the galaxies would have all collapsed to the center of the cluster, and if it was any weaker, the cluster would have dispersed. Hence, Zwicky concluded that by measuring the speeds of many galaxies in the cluster (using the Doppler effect), he could deduce the required gravitational pull and thereby the mass. Much to his amazement, he found that the required mass exceeded by far the visible mass. Zwicky's results met with much skepticism from most of his colleagues, but history was about to prove him right.

All astronomical objects emit radiation at a variety of wavelengths. Only a small part of that radiation is visible to the human eye. Much of it is emitted in the form of ultraviolet, infrared, X-ray, or radio radiation, all of which is completely invisible us, but which can be detected by telescopes either on the ground (e.g., radio telescopes) or in space (e.g., on board ultraviolet, infrared, and X-ray satellites). Observations in these wavelength ranges open for us entirely new windows to the universe. To appreciate the importance of such observations, imagine that of all the colors in the rainbow we could see only the color blue. Everything that is red, yellow, green, and so on would appear completely dark to us. Clearly, our perception of the world would have been incomplete, to say the least. For example, in my absolutely favorite painting, Vermeer's mesmerizing *Girl with a Pearl Earring,* we would see only the crystalline blue turban on the girl's head. In fact, Picasso's melancholic paintings from his Blue Period (between 1900 and 1904) would have been quite clear, but many of those of a less austere mood from the Rose Period (between 1904 and 1905) would have been almost entirely invisible. (Of course, these painters would not have used the same colors had we been able to see only blue.)

Observations with radio telescopes at a wavelength of 21 centimeters have helped to map the outer parts of disklike galaxies. In particular, radio observations reveal the presence of clouds of gas, at distances from the centers of these galaxies that far exceed the radius of the luminous (visible) part. These clouds are found to orbit around the galaxy centers in the same way that the earth orbits the sun. Furthermore, using the Doppler effect, the speeds at which these clouds move in their orbits can be determined. Here, however, came a huge surprise. Suppose that the luminous part of the galaxy represents most of its mass; one would expect that the *farther* the cloud is from the luminous region, the more *slowly* it would be moving. In the same way, the higher the orbit of a satellite above the earth's surface, the slower it moves; similarly, Neptune moves much slower than Venus because it is farther away from the sun. Instead, observations showed that clouds that were extremely far from the center (and way beyond the optically visible part of the galaxy) moved at just the same impatient pace as clouds that were five times closer in. These observations confirmed previous findings by astronomer Horace Babcock at the University of California at Berkeley, and in particular by Vera Rubin of the Carnegie Institution in Washington, D.C., that outlying stars, toward the edge of the visible disk, move just as fast as stars in the main disk. This was a well-deserved success for the young Rubin, who had to overcome prevailing prejudices against women in science at that time (the 1950s and 1960s).

There are only two ways to explain the high speeds of these clouds. We have to conclude either that Newton's law of gravitation breaks down in the circumstances prevailing in the outskirts of galaxies, or that the high orbital speeds are caused by the gravitational attraction of invisible matter. If galaxies were made of only the stuff that we can see, they would not exert a sufficiently strong gravitational pull to keep these speedy clouds orbiting. My friend the Israeli astrophysicist Mordehai ("Motti") Milgrom suggested that the former is happening. Namely, Milgrom proposed that the universal law of gravitation has to be modified when gravity is weaker than some critical value. In spite of some interesting properties, Milgrom's conjecture has not gained many supporters, primarily because it has never been developed into a truly complete "theory." Astronomers therefore have been forced to accept the second possibility: galaxies must contain vast amounts of dark matter that extends far beyond the visible disk of a given galaxy and forms a spherical dark halo around it. The dark halo has to be about ten times (or more) larger in extent than the luminous disk, and it must contain about ten times more matter than the visible matter. This conclusion was reinforced by recent observations of the speed at which Leo I, a distant satellite galaxy to the Milky Way, is moving. Lick Observatory astronomer Dennis Zaritsky has shown that the speeds of about a dozen small galaxies orbiting the Milky Way (including Leo I) indicate a mass more than ten times that of the visible matter. Thus, the luminous matter is a bit like the white foam on the crest of dark massive waves, or the minilights on a dark Christmas tree.

But the dark matter is not confined to individual galaxies. Astronomers repeated the work of Zwicky with modern telescopes, by looking at large clusters of galaxies—collections of a few thousand galaxies, a few million light-years across. As I explained above, from the speeds of the galaxies one can deduce the total mass that is required to prevent the cluster from either collapsing to its center or dispersing completely, with all of its constituents flying apart. In equilibrium, the force of gravity of the matter in the cluster (which is determined by its mass) exactly balances the proneness to scatter in all directions. In this way, careful modern observations confirmed Zwicky's original finding and showed that *more than 90 percent of the mass in clusters is dark.*

The same picture repeats itself on the largest scales on which mass densities have been measured—that of superclusters of galaxies. Superclusters are aggregates of clusters of galaxies, covering a few tens of millions of light-years. Here again, luminous matter constitutes no more than a small percentage of the total mass.

Therefore, there is probably no escape from the conclusion that *most of the matter in the universe is dark.*

When I was a small child, my grandmother took me once to a relatively small circus-type show. In that show, one act in particular left on me an unforgettable impression. It went like this: the stage was at first in complete darkness. Then a small projector turned on, which lit two white-gloved hands that looked as if they were free floating in midair. The hands were doing a variety of gestures and portraying the shapes of various animals. I was absolutely spellbound. When the act finished, all the lights came on and I could finally see how it was performed. A man dressed all in black, with a black mask on his face, was standing in front of a black curtain. The lighting during the performance was such that only his gleaming white gloves could be seen.

The luminous parts of galaxies that are visible to us can be likened to those white gloves; they are being manipulated by the dominant gravitational influence of the dark matter behind (around) them. But the question is, then, what is this dark matter?

Exotica?

In almost every society there are two types of socioeconomic extremes, and they both encounter problems, although of a rather different nature. On one hand, there are those who are so poor that they have to truly struggle to make ends meet, while on the other, there are those who are so wealthy that they have to hire a team of experts to help them decide in what to invest.

The situation concerning the dark matter in astronomy is more similar to the latter. The problem is not that astronomers do not have any idea what the dark matter might be; rather, there are too many dark matter candidates to choose from.

In broad terms, there are three forms that the dark matter could take, and these forms follow precisely the examples I gave at the beginning of this chapter. The dark matter could consist of ordinary but nonluminous matter, such as planets and small stars that did not quite make it to shining on their own, or collapsed dark remnants of massive stars like black holes. A second possibility is neutrinos. A third candidate is some exotic elementary particle that is a relic of the very early universe.

On the scale of individual galaxies, the simplest option, in principle, for a dark matter candidate is stars that do not shine. Oddly, this means either very small stars or the remnants of very massive stars. Stars like the sun shine because a furnace of nuclear reactions produces energy at their centers. However, in stars that are less than about 8 percent of the mass of the sun, the temperature in the core never becomes high enough to ignite the nuclear reactions. Such small stars, known as *brown dwarfs,* never quite make it to a star status, and they remain faint. Of course, normal planets, which are even smaller, also produce very little (if any) light of their own.

Even though brown dwarfs do not generate nuclear energy, they still produce small amounts of radiation due to the release of gravitational energy via slow contraction. Because these objects are relatively cool, most of this low-energy radiation is emitted in the infrared. At present, although a few tens of brown dwarfs and about twenty extrasolar planets have been detected (all in the past few years), it is far from clear whether these objects are numerous enough to account for all the dark matter in the halos of galaxies. In principle, it is possible that when galaxies first form out of primordial gas clouds, brown dwarfs form in huge numbers and continue to roam the dark halos. Armies of brown dwarfs could therefore be holding galaxies together. However, some recent observations cast doubts on this possibility. These observations are based on what are known as *gravitational microlensing events*—the bending of light rays by gravity.

For example, if a brown dwarf in our galaxy's halo crosses our line of sight to a distant star, then the gravity of the brown dwarf acts like a lens, in that it bends the light beams of the distant star. This was one of the predictions of Einstein's theory of general relativity, and it was spectacularly verified during a solar eclipse in 1919. The "lens" temporarily increases the brightness of the distant star. Objects in the halo (such as brown dwarfs or black holes) that can potentially produce such lensing events have fancifully been dubbed MACHOs, for *massive compact halo objects.*

If the entire dark halo is composed of brown dwarfs, then trillions of them should exist. Thus, there is a chance that every once in a while, one of them would cross the line of sight to a distant star and cause a temporary brightening of the star (a microlensing event), lasting for a few days or weeks. Since the amount of bending of the light is larger the higher the mass of the MACHO, more massive MACHOs cause longer-lasting brightening events. The duration of the events can therefore be used to determine the mass of the invisible MACHOs. The idea, first proposed by Princeton University astrophysicist Bohdan Paczyński in 1986, is therefore the following: by monitoring many stars in the two galaxies that are nearest to the Milky Way, the Large and Small Magellanic Clouds (LMC and SMC, respectively), one can hope to detect microlensing events. The number of these events, and the masses deduced for the MACHOs, can then be used to place limits on the total amount of dark matter that can reside in MACHOs. This monitoring experiment is a difficult observation for two reasons: (1) The microlensing events are so rare that literally millions of stars need to be monitored, and (2) one needs a way to distinguish brightenings resulting from microlensing from the temporary brightenings that many variable stars show anyhow. In spite of these difficulties, several teams in the United States, France, and Poland embarked in the early 1990s on the heroic task of finding the MACHOs. The largest of these experiments, led by Charles Alcock of the Lawrence Livermore National Laboratory, was designed to monitor about 10 million stars in the Large Magellanic Clouds and the Small Magellanic Clouds, with each star being observed every two days. At the time this book was written, no fewer than sixteen microlensing events have been discovered (toward the LMC and the SMC), and initially this caused a huge excitement: it looked as if the mystery may have been solved. However, a careful analysis, mainly by a young colleague at the Space Telescope Science Institute, Kailash Sahu, revealed that at least two and possibly most of the observed microlensing events were caused by normal stars in the LMC and the SMC themselves, rather than by

MACHOs in the halo of our own galaxy—that is, stars in the LMC crossed the line of sight to the stars being monitored. Hence, although a firm conclusion will have to wait for further observations, it appears that brown dwarfs may not be the main constituent of dark matter in the halo, after all.

As I noted earlier, besides brown dwarfs, there are other stars that do not shine that could be dark matter candidates. These are remnants of massive stars—black holes. The gravity of black holes is so immense that light cannot escape from them. Thus, in isolation, these objects are completely dark. There are, however, reasons to suspect that this is also not a viable solution for the dark matter in halos. Massive stars, which are the progenitors of black holes, possess strong stellar winds during their lifetimes and they end their lives in gigantic explosions, known as supernova explosions. The winds and the ejecta from these explosions deposit into the interstellar medium (the gas and dust that occupy interstellar space) elements such as oxygen, neon, and silicon, which are the yields of nuclear processing. If black holes were to provide all the dark matter in the halo of our galaxy, then the abundances of these elements would have far exceeded the values determined by observations. Furthermore, unless the masses of these putative black holes exceed ten times the mass of our sun, then the fact that until now no such objects have been detected by the microlensing experiments shows that there just isn't enough mass in black holes to make them a dominant dark matter component. Thus, the only possibility for black holes to play a significant role in dark halos is if they are born from stars that are hundreds of times more massive than the sun. In such cases, no supernova explosion is expected. Rather, the hole can "swallow" most of the yields from nuclear processing, leaving no incriminating evidence for the existence of the massive progenitor. However, stars this massive have never been observed; hence, even their mere existence, let alone the presence in large numbers of their remnant black holes, is highly speculative at this stage.

Even though on the galactic scale much of the dark matter *may* still be ordinary, or baryonic (consisting primarily of protons and neutrons), there are indications that on the larger scales, those of clusters and superclusters, additional forms of dark matter are absolutely required. What can this dark matter be?

One possibility is neutrinos. Neutrinos were originally believed to be massless and hence incapable of producing gravitational effects. However, more recent particle physics theories allow the neutrinos, especially of the muon and tauon variety, to possess a small but finite mass. It is expected that from the big bang, hundreds of millions of neutrinos have been left for every proton. Thus, even if the mass of the neutrino is only one ten-thousandth of the mass of an electron, there could be enough mass in neutrinos to account for all the dark matter.

Neutrinos are extremely difficult to detect. They have no electric charge, and therefore they do not emit any electromagnetic radiation. They also interact extremely weakly with matter. In fact, every second, every square centimeter of our body is bombarded by tens of billions of neutrinos from the sun, and we don't feel anything! This has actually led to a joke about the "neutrino theory of death." If you calculate how much time it would take for one neutrino from the sun to interact with an atom in your body, you find about seventy years; so, the neutrino theory of death says, "That's the one that kills you!" In order to be able to detect neutrinos—from the sun, for example—huge detectors have been constructed. To minimize interference by other subatomic particles, these detectors are placed deep underground; one is in a salt mine in Japan, one in a gold mine in South Dakota, one in a zinc mine in Ohio, one under the Gran Sasso mountains in central Italy, and so on. The depth ensures that most particles are unable to penetrate the earth's crust and reach the detector. The oldest of these experiments, run by physicist Ray Davis Jr., has been operating for more than twenty-five years. It consists of a 100,000-gallon tank of cleaning fluid installed 4,850 feet deep in the Homestake gold mine in Lead, South Dakota. Astrophysicist John Bahcall, of the Institute for Advanced Study in Princeton, has probably done more than any other single person to emphasize the potential significance of these experiments for theories of particle physics in general and for cosmology in particular.

In June 1998, after two years of experiments, a team of Japanese and American researchers led by Y. Totsuka and Y. Fukuda of the University of Tokyo announced results that are consistent with neutrinos having a small mass. This gigantic experiment, called Super-Kamiokande, contains a twelve-story-high water tank filled with fifty thousand tons of exquisitely pure water. The neutrinos are created when high-energy charged particles known as cosmic rays smash into the earth's atmosphere, breaking atoms apart. Billions of neutrinos pass each second through the detector, but only once in a while does one interact with a proton or a neutron in the water, creating a flash of light. Phototubes that cover the walls of the water tank record each event. Unfortunately, the experiment cannot really *determine* the neutrino mass, but rather what minimum value it should have. The minimum mass was found to be about one-thousandth of 1 percent of the mass of the electron.

However, there exist some indirect indications that even if neutrinos have a mass, they *do not* constitute the vast majority of the dark matter. These hints have to do with the formation of structure in the universe.

As I noted earlier, we observe structure on several scales. For example, there are galaxies (on a scale of tens of thousands light-years), clusters of galaxies (on a scale of a few million light-years), and superclusters (on a scale of tens of millions of light-years). Anyone who ever watched a street artist (one of those who draw your portrait) at work knows that there are two ways, in principle, in which such structures might have formed. Some of the artists start by drawing every feature of the face of the sitter, like an eye or the mouth, in great detail, and only later do all the different elements combine to form the complete portrait. Others start with a very broad outline of the entire face, and only later break it up into different features, with the small details appearing last.

The same two possibilities exist for the formation of structure in the universe. The structure might have formed *hierarchically* (or "bottom up"); namely, matter first condensed on subgalactic scales, which later coalesced to form galaxies, which still later formed clusters and eventually superclusters. Or, alternatively, superclusters might have formed first, perhaps in the shape of gigantic cosmic pancakes, with the smaller scales being obtained by successive fragmentations (in a "top-down" process). These two possibilities correspond to two types of dark matter, which have rather different properties.

If the dark matter is hot—that is, its constituent particles are moving frantically at high random speeds in the early universe—then it cannot form the small structures first (bottom up). Any small-scale pattern is, in this case, quickly erased by the random motions. Hence, hot dark matter will inevitably result in a top-down structure formation. On the other hand, cold dark matter, in which the particles move slowly, as in a cold gas, will form the cosmic structure hierarchically. Neutrinos, because of their small mass, are characterized by high random speeds in the early universe and are therefore hot dark matter candidates. However, computer simulations of the formation of structure in the universe produce results that agree much better with observations when a cold dark matter is assumed. In particular, observations indicate that individual galaxies existed, even in the very distant past, when the universe was less than one-tenth its present age, while many huge clusters and superclusters of galaxies did not. Clusters probably formed only during the past 7 billion years or so, and superclusters fairly recently. These observations agree well with the fact that cold dark matter produces the structure in a hierarchical fashion, with the largest clusters and superclusters forming last. In simulations with hot dark matter, on the other hand, individual galaxies form unacceptably late in the life of the universe.

Consequently, it appears that neutrinos are not the dominant dark matter constituent, although they definitely may contribute some fraction of it. The question has therefore been reduced to: What else could this cold dark matter be? Since all other possibilities have been eliminated, this appears to lead to one unavoidable conclusion: *The mass of the universe is dominated by nonbaryonic cold dark matter.*

We know that the dark matter's constituent particles must interact very weakly with matter; otherwise we would have detected them by now. In fact, like the neutrinos, they can interact with matter only through gravity and the weak nuclear force. They must also either be fairly massive (at least fifty to five hundred times the mass of the proton), or be produced at an exceedingly cold state, in order for them to be moving slowly (to be considered *cold* dark matter). Particles of the first kind are known as WIMPs, for *weakly interacting massive particles.* In fact, the MACHOs were whimsically named that way, to contrast them with the WIMPs, which were suggested earlier as dark matter candidates.

But what are the WIMPs?

At the beginning of the chapter, I described the story of the omega minus, a particle predicted (in the early 1960s) to exist on the basis of symmetry alone. In the 1970s, as part of the progress toward unification of the forces, more encompassing symmetries were suggested. Subatomic particles generally fall into two classes according to the value of their quantum mechanical spin. Particles like the electron have a fractional spin (of one-half unit) and are called *fermions.* Particles like the photon have an integer unit spin and are called *bosons.* Very broadly speaking, matter is made of fermions, such as the electron or the proton, while bosons are the carriers of forces, such as the photon and the graviton.

The symmetries suggested in the early 1960s provided classification schemes that grouped fermions and bosons *separately.* In the 1970s, however, more ambitious theories linking the two groups started emerging. These theories became known as supersymmetry (or, affectionately, SUSY) theories. Supersymmetry requires the existence of yet-undiscovered particles (string theories also incorporate SUSY). Some of these particles, which could have been produced copiously in the very early, high-energy universe, might have survived to the present. For example, in SUSY, the familiar particles with integer spin, such as the photon, have "close cousins" that have fractional spin angular momentum. Thus, according to SUSY, there must be twice as many elementary subatomic particles as we have discovered so far, since each particle should have a superpartner. All of the partner particles are supposed to be uncharged electrically, to interact very weakly with ordinary matter, and to be rather massive. Hence, these are perfect candidates to be the missing WIMPs. Most of these particles are extremely unstable and fall apart in no time, but the leading WIMP candidates are stable. These are the lightest SUSY particles, known as *neutralinos*. According to the theory, vast amounts of neutralinos should have been created in the early universe.

I should emphasize that we do not know yet if such particles really exist. To date, no neutralino or any other WIMP has been detected experimentally. However, several experiments designed to search for neutralinos are under way. Some of these experiments use powerful particle accelerators in an attempt to produce neutralinos directly. Others use sensitive underground detectors that search for the rare flashes of light produced by interactions that neutralinos in the Galaxy's halo might produce, once intercepted by the earth. At the time of writing, the Large Hadron Collider, the world's most powerful particle accelerator, being built in Geneva (by the European Organization for Nuclear Research, or CERN), is about (within the next decade) to achieve the energy range that is necessary to test the predictions of SUSY.

As I noted above, cold dark matter particles could also be of low mass, if they are produced in an exceedingly cold state. An example of such an exotic particle, predicted to exist by particle physicists Steven Weinberg and Frank Wilczek, is the axion. Unlike the neutralinos, axions are actually predicted to be astoundingly light (about one-trillionth the mass of the electron). In principle, axions can be transformed into radiation (microwave photons) using a strong magnetic field. Early experimental searches for axions failed to find a particle with the anticipated properties. At the time of writing, an experiment at Livermore, which represents a large collaboration of U.S. institutions, is reaching a critical phase in its search for axions, with a promise to produce conclusive results within the next five years.

In October 1998, a team working at the Gran Sasso Laboratory under the Apennine Mountains in Italy made the tantalizing announcement of "a possible WIMP candidate." Analyzing two years worth of data, the team (dubbed DAMA, for "dark matter") found more low-energy flashes in the summer than they did in the winter. This is what might be expected from WIMPs: in the summer, the earth is moving around the sun in about the same direction as the sun's motion around the center of our galaxy. Thus, the two motions add up, and the earth is moving faster through the presumed halo of WIMPs than it does during the winter. Many scientists are still skeptical, however, about this result, since all the potential sources of error have not been fully identified.

So, where does all this leave us? Most astrophysicists are convinced that at least 90 percent of the mass of the universe is dark. Opinions start to vary more when one considers what the dark matter could be. Some, like cosmologist Joseph Silk at Berkeley, would probably still prefer all the dark matter to be baryonic, like MACHOs. Others point out that it is highly unlikely that baryons alone will provide enough dark matter for clusters of galaxies (I will return to this topic later), and advocate cold dark matter in the form of exotic particles like neutralinos. It should be realized, however, that this is not some futile pseudophilosophical debate. Experiments searching for neutralinos (and axions) will soon, it is hoped, be able to tell us whether such particles exist. Surprises are nevertheless still possible, as the searchers for proton decays have shown. For example, as I mentioned in chapter 3, the "minimal version" of a GUT theory that unifies the forces predicted that protons should decay to a positron and a particle called pi zero in about 10^{30} years. However, the existing experimental results from three experiments already show that protons do not decay to these particles for at least 10^{33} years. This shows that the underlying grand unified theory is surely different from the original minimal model. Similar complications may push the discovery of neutralinos into the more distant future. A total failure to discover substantial amounts of dark matter in any form may force physicists to consider suggestions for a modified nature of gravity more seriously. It may be some time before we can declare the case of the missing matter closed.

One may ask what the implications are of the existence of cold dark matter for the beauty of the physical theory of the universe. Can we use the words of songwriter/singer Paul Simon: "Hello darkness, my old friend"? On one hand, clearly the discovery of exotic cold dark matter particles will record a remarkable success for the predictions of the underlying symmetry. Furthermore, if these exotic particles are the main constituents of the matter in the universe, this takes the applicability of the Copernican principle yet another giant step forward. Not only is our position in the universe nonspecial, but even the stuff we are made of, ordinary baryonic matter, represents only a small percentage of the matter in the universe. While from a global point of view this makes it more appropriate to call the matter *we* are made of "exotic," this still does not make us particularly special, since all the matter we *see* is made of the same stuff. On the other hand, however, there is one aspect of dark matter that appears on the face of it to be somewhat worrisome from a beauty perspective. If, for example, some nonnegligible fraction of the dark matter on the galactic scale is provided by MACHOs and only the rest is exotic cold dark matter like WIMPs, this seems like a slap in the face of reductionism. Why should there be two types of dark matter?

Could this be a first crack in the otherwise perfect canvas of beauty I have painted so far? Not necessarily.

One has to realize that the aesthetic principles are generally applied to the *fundamental* aspects of the theory and to its cornerstone *idea* and not to the more peripheral details. There is nothing particularly fundamental about the individual constituents of dark matter—although, as the next chapter will show, there could be something fundamental about the *total mass* in dark matter.

The identification of what is fundamental is not always easy. The history of science is full of examples of concepts and entities once considered absolutely fundamental at certain periods of time, but that have been knocked off their pedestal of fundamental status at later times as a deeper understanding of the workings of nature emerged.

To Plato, for example, the elements earth, fire, air, and water were fundamental; to Copernicus, the planets were the "heavenly spheres." Most of the subatomic particles that were considered elementary in the early days of particle physics were later shown to merely represent higher energy states of more fundamental entities. Even protons and neutrons, once considered to be the *basic* building blocks of matter, were later found to be composed of yet more basic particles—the quarks.

As a result of this ambiguity, it can definitely happen, and indeed it has happened, that the aesthetic principles would be applied to the wrong entities altogether or only to a subset of all the relevant entities. An example of the former is the Titius-Bode law; an example of the latter are Einstein's endeavors to unify the electromagnetic and gravitational forces.

The Titius-Bode law was an attempt by eighteenth-century astronomers to identify some underlying harmony and regularity in the structure of the solar system. In 1766, the German astronomer Johann Titius came up with a simple mathematical formula that predicted relatively accurately the radii of the orbits of the then known planets: Mercury, Venus, Earth, Mars, Jupiter, and Saturn. This formula was published and popularized by the astronomer Johann Bode in 1772.

Unfortunately, the orbit of Uranus, discovered in 1781, already deviated somewhat from the prediction, and in predicting the orbit of Neptune the formula failed miserably. At present, few astronomers treat this "law" as anything more than a mathematical curiosity, although the possibility still exists that it holds some information about the formation of the solar system. At any rate, the point is that there is nothing particularly fundamental about planetary orbits, and therefore there need not be some beautifully unifying theory that explains them.

I have already mentioned the fact that Einstein's attempts in the unification of the forces mistakenly concentrated only on the gravitational and electromagnetic interactions, ignoring the strong and weak nuclear interactions. This was an oversight based on Einstein's greater familiarity with the former two forces, which led to him assigning to them a more fundamental role in his mind.

Returning to the question of whether the existence of several kinds of dark matter should be regarded as a violation of reductionism, the above examples show that such a conclusion would be premature at best. We first have to be able to assess better the fundamentality of the dark matter. To this goal, we will now examine the role that dark matter plays in determining the ultimate fate of the universe.

5

Flat Is Beautiful

In Shakespearean tragedies we always know when the story ends. These plays convey such a strong sense of closure that we are not left with much curiosity about the future (partly because all the main characters are dead!). When discussing the evolution of the universe, however, as in our own personal lives, we are at least as curious about the future as we are about the past. So far I have only discussed the past. In doing so, I have given several examples of how the basic elements of beauty—symmetry, simplicity, and the Copernican principle—have become delicately interwoven in a theory for the early universe and for its subsequent evolution. This theory has enjoyed considerable success in explaining the universe we observe today. The question that arises now, however, is: *What will the universe do in the future?* As we shall soon see, perhaps never in the history of physics have aesthetic arguments played a more dominant role than in the attempts to answer this question.

The most important factor in determining the ultimate fate of the universe is gravity. Gravity can (in principle) reverse the current expansion and lead to a universal collapse. What is usually not fully appreciated is the fact that in his theory of universal gravitation, Newton achieved the first unification of forces. He recognized that the same force that makes apples fall on Earth also holds the moon in its orbit around the earth and the planets in their orbits around the sun. Newton was thus the first to use the aesthetic principles of symmetry and reductionism to construct a universal theory, precisely in the same way that these principles are used in modern particle physics theories. It is truly fascinating to follow briefly the chain of events and the line of thinking that led to this remarkable theory of gravitation.

Newton arrived in Cambridge in 1661, a time in which England was returning to a more secular life, following a rather austere puritanical period. During his years at Trinity College, he had to help finance his education by working as a part-time servant to wealthier students. In spite of his numerous duties (which, by the way, included emptying chamber pots), he found time to fill entire notebooks with observations of natural phenomena, such as the properties of light. In 1664, however, sleep deprivation brought him to a state of complete exhaustion. Nevertheless, he managed to complete his bachelor of arts degree and was about to start his studies toward a master's degree when news of the Great Plague in London reached Cambridge. Death was taking a toll of more than ten thousand people a week in the densely populated and poorly sanitized city. Even before Cambridge University was officially closed, Newton decided in 1665 to return to his home in Woolsthorpe, Lincolnshire. It was there that on one fine evening the legendary incident with an apple occurred. William Stukeley, in his *Memoirs of Sir Isaac Newton's Life* (published in 1752), gives the following description of how the story with the apple became known:

> After dinner [on April 15, 1726], the weather being warm, we went into the garden and drank tea, under the shade of some appletrees, only he [Newton] and myself. Amidst other discourse, he told me he was just in the same situation, as when formerly, the notion of gravitation came into his mind. It was occasion'd by the fall of an apple, as he sat in a contemplative mood. Why should that apple always descend perpendicularly to the ground, thought he to himself. Why should it not go sideways or upwards, but constantly to the earth's centre? Assuredly, the reason is, that the earth draws it. There must be a drawing power in

> the matter: and the sum of the drawing power in the matter of the earth must be in the earth's centre, not in any side of the earth. Therefore does this apple fall perpendicularly, or towards the centre. If matter draws matter, it must be in proportion of its quantity. Therefore the apple draws the earth, as well as the earth draws the apple. That there is a power, like that we here call gravity, which extends its self thro' the universe.

It is a mistake to think that Newton "discovered" gravity. (In a satirical television program I saw many years ago, they were ridiculing "scientific" programs, and said: "Newton discovered gravity in 1665, but actually gravity *existed* four hundred years before Newton!") Apples had surely been seen to fall before Newton's time, and the cause for their falling was correctly attributed to a mysterious force of attraction possessed by the earth, which had been named *gravity*. Newton's great "unifying" contribution consisted of showing that this gravity, which until then was assumed to be a peculiar property associated with the earth, was in fact a universal property of matter. He transformed an oddity into a *beautiful* theory. It is instructive to follow Newton's original train of thought. An apple certainly falls down from a tree that is fifty yards high. In fact, it would fall even from the highest mountaintop, and probably from much higher. It is conceivable that the farther you go from the earth the weaker the attraction that the apple feels, but is there a distance at which the attraction vanishes? The nearest astronomical body to the earth is the moon, some 240,000 miles away. Would an apple thrown from the moon also reach the earth? But maybe the moon itself has an attractive force, in which case, since the apple would be so much closer to the moon, it would probably fall on the moon. Now, wait a minute! Not only apples fall to the ground. What is true of the apple is true of all matter, large or small. The moon itself is a very large body; does the earth exert any gravitational pull on the moon? It is true that the moon is hundreds of thousands of miles away, but it is also a huge body, and perhaps the amount of matter is somehow related to the power of attraction. However, if the earth attracts the moon, why isn't the moon falling to the earth? The moon is actually not at rest, but traveling at a very large speed, because it circles the earth in a month. Now, if the earth was not there, the moon would travel along a straight line. Because of the earth's attraction, the moon's path is continuously being curved, and it is probably elliptical, just as Kepler found for the planets. So, the only reason that the moon does not fall to the earth is on account of its motion.

From this type of deductive thinking, coupled to the use of Kepler's observations of the motions of planets around the sun, Newton derived his universal law of gravitation. He then calculated the attractive force that the earth exerts on the moon, and compared it to the force the earth exerts on an apple. He writes: "I compared the force necessary to keep the moon in its orb with the force of gravity at the surface of the earth, and found them answer pretty nearly." This is truly a magnificent example of a reductionistic approach and its experimental test.

Returning now to the future behavior of the universe, we have to apply precisely the same laws. Examine, for example, what happens to an object thrown into the air on earth. If we toss an apple up, it reaches some maximum height where it stops momentarily, and then it returns to our hands. This happens, of course, because of the force of gravity. The force of gravity is sufficient in this case to brake and decelerate the moving apple to the point where it reverses its velocity. By now, however, we have become equally familiar with pictures of rockets that leave the earth on no-return journeys. The question is, then: What is the difference? Simply, in the latter example, the rockets reach such speeds that their kinetic energies of motion exceed the gravitational energy. In such a case, the earth's gravity is incapable of stopping the motion.

The situation with the universe as a whole is similar, in principle, to that in the above examples. If the kinetic energy that is involved in the expansion of the universe is *smaller* than the gravitational energy due to all the mass within the universe, then the expansion will come to a halt, and the universe will start contracting, eventually leading to a collapse in a *big crunch.* The final stages of this collapse will mirror the big bang in reverse motion. If, on the other hand, the kinetic energy is *larger* than the gravitational, the expansion will continue forever, with the speed of expansion never reaching zero. Galaxies will eventually exhaust all of their gas reservoirs for forming new stars, and the old ones will fade away and die. The universe will cool down continuously. Eventually, even the protons will decay, as the universe will approach a cold death. In the borderline between the fate of a hot inferno or a big chill, when the kinetic energy *exactly equals* the gravitational energy, the expansion still proceeds forever; however, the expansion speed approaches zero as time progresses. The question is then: Which of these three possibilities represents our universe?

The gravitational energy is larger the higher the mass density. For example, a neutron star is a very compact star that is obtained from the collapse of the core of a very massive star. Neutron stars have a mass similar to that of the sun, but a radius that is seventy thousand times smaller. Consequently, gravity near the surface of a neutron star is about 5 billion times stronger than near the surface of the sun.

The gravitational energy of the universe is *precisely equal* to its kinetic energy for a particular value of the mass density in the universe. This value, which separates eternal expansion from eventual contraction, is called the *critical density.* If the density in the universe is higher than the critical density, gravity will prevail—the expansion will stop and contraction will ensue. If, on the other hand, the density is lower than critical, the universe will continue in perpetual expansion. It is customary to denote the ratio of the actual density to the critical density by the capital Greek letter omega (Ω). Thus, eventual contraction and a big crunch corresponds to a value of omega that is larger than one. Expansion ad infinitum is represented by an omega that is smaller than one, and a density that is exactly equal to the critical density corresponds to a value of omega *exactly equal to one.* Remember that for an omega of one, the universe also expands forever, but with a speed that slowly approaches zero.

Therefore, in order to answer the question about the ultimate fate of the universe (in terms of continuing expansion and a "big chill" versus a "big crunch"), all we need to determine is the *present* value of omega* or, in other words, to determine whether the density of mass in our universe is higher than, lower than, or equal to the critical value.

If you think about it for a moment, this is a remarkably fortunate situation, both from a physics perspective and from a psychological point of view. By determining observationally the value of one quantity, we will be able to know the fate of the universe as a whole. The universe, in this sense, is thus quite simple (as required for beauty). Determining the value of omega is thus almost like a fulfillment of Shakespeare's wish in *Henry IV:* "O God! that one might read the book of fate."

Fire or Ice?

So, how much is the critical density, and is the density in the universe higher or lower than that? The critical density is about 1 atom in a volume of 8 cubic feet (or 5 atoms in 1 cubic meter). This is about 10^{26} times more rarefied than air and about 100 billion times more dilute than what we would consider to be an extremely good vacuum. Thus, at first glance it may seem as if it is not too difficult for the density in the universe to exceed this value. However, once we realize the vastness of the nearly empty spaces between galaxies, the answer becomes much less obvious. In fact, it turns out that if we were to disperse all the luminous matter in galaxies uniformly

---

*Later in this chapter we shall see that the situation is actually somewhat more complicated. However, the above description certainly represented the general thinking until a few years ago.

in the universe, the density would still be about 100 times *smaller* than the critical density (corresponding to a value of omega of less than 0.01). Thus, if this were to represent all the matter that exists in the universe, then there would have been no doubt that the universe would go on expanding forever. However, we remember from chapter 4 that in fact the universe contains huge amounts of dark matter, which is the dominant source of gravity. In principle, therefore, omega could be even larger than 1.0. How does one go about determining the value of omega, or the mass density in the universe? The methods employed are very similar to that used by Hannibal, the great Carthaginian general, before his battle with the Romans in Cannae, in 216 B.C. The story has it that he managed to determine quite accurately the size of the Roman army, from information on the food supplies sold to this army. Similarly, the attempts to measure the cosmic mass density rely on the observable effects that this density can produce, mainly through its gravitational attraction.

There are several independent methods that have been used to determine omega, and I will describe here only a few of them. One such method has already been mentioned in chapter 4. From the speeds of clouds around the centers of individual galaxies and the speeds of galaxies in clusters and superclusters, astronomers have established that the dark matter overweighs the luminous matter by a factor of about 10 or more. Thus, *the value of omega inferred from dynamics in clusters and superclusters is about 0.2 to 0.3.* Does this, however, complete the inventory of dark matter? Not necessarily. In principle at least, there could still be vast amounts of dark matter spread throughout the intergalactic-intercluster space. The existence of such matter would generally not show up in the motions of galaxies in clusters, and certainly not in the motions of clouds inside individual galaxies. So how can such matter be detected or its existence inferred? One way is through large-scale motions. The point is the following: in spite of the fact that on the *largest* scales the universe is homogeneous and isotropic, on smaller scales it clearly is not, as manifested by the mere existence of galaxies, galaxy clusters, and superclusters. Thus, for example, on a scale of a few tens to hundreds of millions of light-years, our galaxy is "falling" toward the Virgo cluster of galaxies, because of the greater gravitational attraction from that direction. In fact, the Virgo cluster itself is being pulled toward yet higher-mass concentrations. By identifying coordinated motions of large numbers of galaxies on scales of a few hundred million light-years, one can therefore detect the presence of dense regions in the universe and in some cases determine the amount of mass involved in them. This type of measurement (of speeds of galaxies) led to the discovery of the *Great Attractor*—a large, unseen mass concentration toward which the Milky Way, together with a few other hundreds of neighboring galaxies, are being pulled. The name "Great Attractor" was coined by the astronomer Alan Dressler of Carnegie Observatories, who, together with Sandra Faber, Donald Lyden-Bell, and four other astronomers, discovered this coordinated motion (a nice description of this discovery can be found in Dressler's book *Voyage to the Great Attractor*). It is presently not entirely clear what value omega would reach when such great attractors are included, but the value could certainly be somewhat higher than 0.2, maybe 0.3 to 0.4.

Since every astronomical measurement involves some uncertainty, and in particular since most of the methods for the determination of omega are indirect, it is important to have other methods to be able to compare the results.

An excellent clue to the amount of ordinary (baryonic) matter comes from the cosmic abundances of some very ordinary light elements. Particularly useful in this respect is the heavier isotope of hydrogen called deuterium. Different isotopes of a given atom have the same number of protons in their nucleus, but a different number of neutrons. Hydrogen, for example, has only one proton in its nucleus, while deuterium (sometimes referred to as heavy hydrogen) has one proton and one neutron. Deuterium's usefulness stems from its extreme fragility. At high densities and at temperatures merely exceeding about 1 million Kelvin (less than one-tenth the temperature at the center of our sun), deuterium suffers sufficient collisions that it is effectively destroyed by being converted into helium. Consequently, evolved stars do not contain any deuterium.

Deuterium was formed in the first three minutes following the big bang, from the interaction of free protons with free neutrons. Since then, the deuterium abundance in the universe is only decreasing with time, since deuterium is being destroyed inside stars. The reason that the deuterium abundance is so important is the following: This abundance depends crucially on *the density of ordinary matter* (baryons). If the density of baryons is too high, then essentially all of the deuterium is destroyed in the early universe, by conversion to helium. If, on the other hand, the density of baryons is too low, then deuterium is overproduced. Determinations of the deuterium abundance thus place limits on the density of ordinary matter *in any form.* Therefore, the deuterium abundance serves as some sort of "baryometer," a name coined by cosmologists Dave Schramm and Mike Turner, a measurement tool for the density of baryonic matter.

There are two ways, in principle, in which you might attempt to determine the deuterium abundance in the early universe. For example, imagine that there is a sealed room with a few people inside and at some point you are asked to determine how much oxygen there was in the room *to begin with.* Clearly, your preferred method would be to perform your measurement of the oxygen as short a time as possible after the experiment was begun, since then you would be able to obtain a pretty good estimate. If, however, you are forced to make your first measurement after the experiment has been running for some time already, then all is not lost. What you might do in this case is take two measurements separated by some fixed amount of time. From the difference between the two measurements you could then determine the rate at which oxygen has been consumed, and then knowing when the experiment started you could deduce the initial amount of oxygen. The same two approaches have been applied to the deuterium abundance. Deuterium is detected in interstellar clouds (which are sufficiently cold and dilute to avoid deuterium destruction) and in places like the atmosphere of the planet Jupiter. The important point is that the abundance of deuterium in Jupiter reveals its abundance in the interstellar gas from which our solar system formed, some 4.6 billion years ago. On the other hand, the abundance in local interstellar clouds samples material some of which has been processed through the hot cores of stars that lived and died during the Galaxy's existence. Thus, a comparison of the two values gives an estimate of the *rate of destruction* of deuterium. Once the rate of destruction is known, we can estimate how much deuterium was synthesized in the first three minutes or so of the universe's lifetime. Better yet, by analogy to the first method described above, during 1998, observations by astronomers David Tytler from San Diego and Scott Burles from Chicago yielded the deuterium abundance in some very distant clouds. These clouds are composed of nearly pristine material, representing the universe when it was less than one-fifth its present age. Hence the big bang abundance of deuterium has been established quite accurately. This abundance implies a density of ordinary matter that is about 5 percent of the critical density, and it absolutely cannot exceed one-tenth of the critical density. Thus, if all the matter (including dark) in the universe were ordinary, rather than consisting of some exotic particles, *then omega would not have exceeded 0.1* (and most likely would have been only about 0.05).

The deuterium abundance determines only the density of *baryonic* matter. There exist, however, other methods to determine the density of all matter, including exotic. One such method relies on "weighing" ancient clusters of galaxies. Somewhat paradoxically, perhaps, if you find that distant, very old clusters of galaxies are very massive, then this is a strong indication that the density in the universe is rather *low*. To understand why this is the case, examine the following example from everyday life. Imagine that you value very much the idea of giving presents at holidays and that you are contemplating a strategy for the accumulation of Christmas, Hanukkah, or Kwanza presents. If you think that your current economic situation is not likely to change much, then you could accumulate the presents continuously over the entire year. If, on the other hand, you have reasons to suspect that your economic situation may worsen in the future, then it may be a good idea to accumulate the presents while you still can, relatively early on. The universe operates somewhat analogously, when it comes to building up the mass of clusters.

If the universe is dense (omega of one or larger), then clusters of galaxies are expected to grow in mass *continuously,* since there is always more mass around for clusters to accrete or merge with. Consequently, one does not expect in such a case to find very massive clusters (as massive as one sees today) already in place in the distant past. On the other hand, if the density is relatively low, then compared to the early, relatively denser universe, the subsequent rate of growth of clusters of galaxies is *slowed down* considerably. Thus, in such a case, some clusters that are as massive as the ones we see at present can exist relatively early on.

Recent observations of a few distant clusters, and in particular a study conducted by young colleagues at my institute—Megan Donahue, Mark Voit, and their collaborators—reveal that they are surprisingly massive. One distant cluster, known by the unattractive name MS1054 – 0321, has been nicknamed "a nine-hundred-pound gorilla," because it is at least as massive as present-day clusters, weighing about a few thousand Milky Way galaxies. This cluster was formed when the universe was half its present age. The mass of the clusters can be determined, for example, by measuring (using X-ray observations) the temperature of the hot gas that fills the space between the galaxies. The hotter the gas is found to be, the more massive the cluster needs to be so that its gravitational attraction can prevent the gas from escaping into space, like steam from a hot kettle. From the masses deduced for a sample of such distant clusters, it has been inferred that omega cannot exceed about 0.3 to 0.4.

We therefore see that a number of different methods give a value of omega of about 0.3 to 0.4. Many cosmologists were surprised by this value, because they had other *expectations* for the value of omega. The universe appeared in some sense to be an underachiever. I will describe now these expectations in some detail, since they provide a magnificent example of the fundamental role of aesthetics in physical theories.

Beauty Reigns?

One of the requirements of any true theory in physics is its ability to make *predictions,* which can then be tested by experiments or observations. This is in fact one of the main ways to ensure that progress toward an ultimate, beautiful theory can be achieved. The process is somewhat similar to that of natural selection in the evolution of life on Earth. Through a series of false starts and blind alleys, physical theories are eventually guided in the right direction by a combination of encounters with experimental tests and with theoretical developments. The latter are based partly on those experimental tests, partly on progress in mathematics, and finally on aesthetic intuition.

The situation concerning the value of omega is quite unique, in that even before a theory that *predicted* its value existed, cosmologists developed a very biased expectation, *based entirely on aesthetic arguments.*

In order to understand the argument, consider a tightrope artist, balanced very precisely on one foot. While this represents a state of equilibrium, which in principle could continue for a long time, it is an *unstable* equilibrium; namely, the smallest push would result in a catastrophic departure from equilibrium and a disastrous fall. When the value of omega equals 1.0, this represents a state of unstable equilibrium for omega. That is, if at any time in the past omega was *exactly* equal to 1.0, *it would remain equal to 1.0 forever.* If, however, omega differed even slightly from 1.0 at any instant following the big bang, then *its deviation from 1.0 would be amplified extremely rapidly with time.* For example, if omega were equal to 0.5, then the rapid expansion and the dilution of matter would have resulted in gravity fighting a losing battle, with its effects less and less noticeable. The value of omega would have dropped to 0.25 by the time the universe had merely doubled its size.

Therefore, a universe starting with a value for omega of less than one will expand quickly, thus reducing the value of omega substantially. Such a universe will spend most of its existence at an omega of very nearly zero.

As we saw in the previous section, several methods for the determination of omega indicate a present value of 0.3 to 0.4. But the universe today is extremely old—in fact, about 14 billion years old. Suppose that omega started with a value that was not precisely equal to 1.0, but somewhat smaller. One may ask: How close did it have to be to 1.0 when the universe was, say, only one second old for it to reach 0.3 to 0.4 today, rather than being essentially zero for billions of years already? This is similar to asking how close to perfect equilibrium the tightrope artist had to be for us to still find him on the rope many years later, rather than on the safety net underneath. The value could not have differed from 1.0 by more than one part in 10^{15} (namely, it had to be 0.999999999999999). Thus, for omega to be different from 1.0 in the early universe and yet to be consistent with a value of 0.3 to 0.4 in the present, the early universe had to be incredibly *fine-tuned.* The kinetic energy of the expansion had to be equal to the gravitational energy to a fantastic precision, *but not exactly.* Furthermore, a universe that starts with an omega different from 1.0 either expands quickly (if omega was initially less than 1.0) to a point where the pull of gravity can be neglected altogether (omega of zero), or recollapses in a hurry to a big crunch (if omega was initially larger than 1.0). Yet here we are finding the deviation of omega from 1.0 to be rather modest. Omega is 0.3 to 0.4, rather than zero or infinity—the much more likely values resulting from rapid expansion or recollapse (respectively). This is a bit like finding the tightrope artist still in midair. For us to be able to "catch" omega while it is so close to 1.0 would mean that we are living at some very special time. However, both the necessary fine-tuning and the requirement for an anthropocentric time violate severely my generalized definition of the Copernican principle and are therefore extremely ugly. A way to avoid this requirement for fine-tuning and a violation of the Copernican principle is to assume that *omega is actually exactly equal to 1.0* and that the value of 0.3 to 0.4 inferred by several methods merely represents an inability of these methods so far to uncover all the dark matter that exists.*

As we shall see shortly, there does exist now a theory that *predicts* omega to be exactly equal to one, but even before this theory was developed, physicists expressed a strong prejudice favoring omega equals one, simply on the basis of aesthetics.

This provides a singular example of the deep belief shared by many physicists that inclusive theories of the universe *must* be beautiful. One should realize that there is absolutely nothing in the laws of physics that, for example, *forbids* fine-tuning or a violation of the Copernican principle. Yet most physicists feel that this should not be allowed. For example, my good friend and colleague the cosmologist David Schramm, from the University of Chicago, wrote in a paper in 1984 (with Katherine Freese): "On the basis of 'simplicity' we believe that a non-baryonic universe with Omega larger than 0.15 should satisfy Omega equals to 1." Tragically, Dave's brilliant career and his own pursuit of beauty were cut short when his personal plane crashed in Colorado in 1997.

The above prejudice deserves some further discussion. In the

---

*In the next chapter, I will discuss a modification of this conclusion, based on recent findings.

history of science, there certainly have been instances in which the entire scientific community, some parts of it, or particular individuals within it were influenced by prejudices. A good example is provided by the dominance of the Aristotelian-Ptolemaic geocentric model for the solar system, which continued to provide the intellectual basis for all models of the universe for nearly thirteen centuries. It is partly because of the fact that this prejudice was so deeply rooted that we talk about a "Copernican revolution" when referring to the heliocentric model.

Even the best scientific minds have been known to succumb to prejudices. For example, Einstein felt extremely uneasy because of the threat that quantum mechanics posed to one of the fundamental properties of classical physics—namely, determinism. In classical physics, the evolution of any physical system, and of the cosmos as a whole, can be predicted with complete certainty and precision. Quantum mechanics changed all that, by asserting that only *probabilities* for certain events can be determined. Einstein's prejudice was that there should be a precise one-to-one correspondence between the physical theory and the reality that this theory describes, and quantum mechanics challenged this notion. Consequently, Einstein never really accepted quantum mechanics, as expressed by his famous dictum: "I shall never believe that God plays dice with the world."

Most of the examples of prejudices, like the ones above, are of prejudices related to particular models, particular theories, or the nature of particular objects. The prejudice about the beauty of cosmological models is quite unique in this respect, in that it is related to a basic property that *any* model should have. Since this prejudice was originally not really anchored in any particular insight but rather in a belief that was based on past experience, one could almost argue that it is in some sense closer to religion than to science. Unlike in religion, however, *the ultimate test of the validity of this prejudice will come from precise observations and experiments.*

Of the basic forces of nature, gravity will have the final say in determining the global fate of the universe. When one discusses gravity, two names stand out in the history of science: Newton and Einstein.

Einstein

Sotheby's auction room is crowded to capacity. In the front row sits an old man with a bushy mustache and disheveled gray hair.

AUCTIONEER: The next item is an unusual object from the seventeenth century. It is the apple that fell on Newton's head, and it has been miraculously preserved.

[*The auctioneer ignores the surprised reaction of the audience and continues.*]

AUCTIONEER: I would like to start the bidding at ten thousand dollars.

Curtain. The curtain goes up again some time later. The scene has now changed; the auction room is empty and quite dark. Only the gray-haired man is still sitting in the front row, now holding in his hand a strange-looking apple, and examining it closely. He notices a number of spots on the surface of the apple, and he takes out of his vest pocket a magnifying glass to study them. You can hear that, for a moment, he stops breathing in astonishment, as he looks at the spots through the magnifying glass.

OLD MAN [*to himself*]: How strange, these spots do not all look the same. Some look like small ellipses, others have spiderlike or water sprinklerlike shapes.

[*He now concentrates on one such elliptical spot and its vicinity. Suddenly, he drops the apple abruptly, visibly shaken.*]

OLD MAN [*in an amazed voice*]: Good Lord, they are moving!

[*He picks up the apple with some hesitation and resumes observing the area around one spot carefully.*]

OLD MAN [*to himself*]: It is clear that all the spots around the one I am looking at are moving slowly away from it.

[*He decides to examine a different spot.*]

OLD MAN: Everything is moving away from this one, too.

[*He looks at a few more.*]

OLD MAN: But this is true for all the spots! They are all moving away from each other.

[*After a short pause.*]

OLD MAN: I don't believe my eyes; this apple is expanding.

[*Indeed, it can now be clearly seen that the apple is slowly expanding. The old man takes a piece of paper out of his pocket and starts writing frantically.*]

OLD MAN: [*reciting aloud what he is writing*]: In fact, the spots do not move at all on the skin of the apple, they only appear to be moving away from each other because the apple is expanding. If I were a two-dimensional creature living on one of these spots, I would see every other spot moving away from me. This is really interesting.

[*The apple has by now reached the dimensions of a large watermelon. Strangely enough, the spots themselves do not grow at all; only the distances between them are growing. The old man appears to be making some measurements with his fingers.*]

OLD MAN: [*murmuring to himself*]: Suppose I choose this spot. [*He points to a spot.*] I can find a spot that is one inch away from it on the skin, and one that is two inches away. Now I'll wait a bit for the skin to expand.

[*He sits down, while observing the expanding apple.*]

OLD MAN: [*getting up again and measuring*]: Now the spot that was one inch away from my [*emphasizes "my"*] spot is about two inches away, and the one that was two inches away originally is now about four inches away. This is clear because every distance approximately doubled due to of the expansion.

[*He sits down again, clearly absorbed in some thoughts.*]

OLD MAN: [*getting up, as if he has reached some conclusion*]: But this means that from my [*emphasizes "my"*] spot, I would have seen the farther spot moving away twice as fast as the nearer spot, since it moved by about two inches, while the nearer one only moved by about one inch at the same time. [*He pauses for a minute.*] Maybe my original equations were right after all!

[*The apple has by now grown so large that the old man has to move closer and closer to the wall. Finally, realizing that there is hardly any room left, he flees through the door.*]

Nature Never Makes Any Blunders

The entire discussion of the fate of the universe has centered so far on the value of omega, or the density of matter. Until 1998, most astronomers believed that a determination of this one parameter would reveal the ultimate destiny. However, life is never as simple as one would like, and there exists an extra twist to the description presented in the previous sections. This complication dates back to a paper written by Einstein during World War I. In Einstein's first attempt to apply his general theory of relativity to the entire universe, he realized that there was a problem. The prevailing assumption at the time was that the universe is static—that is, eternal and neither expanding nor contracting. Yet the gravitational field equations did not allow any static solution. This is easy to understand: if galaxies are distributed evenly in space, then clearly they will collapse toward each other under their mutual gravitational attractions. Since Einstein was convinced that the universe is static (because of the war he did not hear that the astronomer Vesto Slipher had already found indications that the universe was expanding), he looked for ways to prevent "his universe" from collapsing. He achieved this by introducing an additional term in his equations, a kind of repulsive force that is most important for objects at large distances from each other. Unlike gravity, which becomes four times weaker when the distance between two particles is doubled, this repulsive force becomes twice as strong. This "cosmological repulsive force" was designed in such a way so as to stabilize the entire universe against gravitational collapse, while its effects at distances smaller than the solar system are completely negligible. The force term involved one parameter for which arbitrary values could be assigned, which became known as the *cosmological constant* and was denoted by the capital Greek letter lambda (Λ). Thus, Einstein basically invented an antigravity that could hold the entire universe under its own weight. Once it became clear that the universe is in fact not static, and the expansion of the universe was definitively established by Hubble, Einstein came to regret the introduction of this additional cosmological constant. Upon retracting it in 1931, he called it the "biggest blunder of his life" (although in the actual paper, written with Willem de Sitter, he uses a much less extreme language: "It now appears that in the dynamical case this end can be reached without the introduction of lambda").

However, as is often the case, even the "blunders" of some truly great individuals turn out to be extremely interesting. The first point to note is that there is nothing in Einstein's theory to *prevent* the existence, in principle, of a cosmological constant. As I explained before, simplicity, as an essential element of a beautiful theory, is meant to be understood as simplicity of the *central idea,* not of the number of terms in the equations.

Einstein's general relativity is based on an underlying *symmetry,* which states that the laws of nature remain and look the same whether we are in a laboratory at rest, in an accelerating rocket, or on a merry-go-round. Thus, we use precisely the same laws on top of our rotating Earth as we would on the surface of a collapsing star, for example. The important point is that the addition of the term involving the cosmological constant does not violate this basic symmetry of general relativity.

After being discarded by Einstein in 1931 and ignored by cosmologists for years, the cosmological constant started to reappear in the physics literature. In 1967, for example, the great Russian cosmologist Yakov B. Zeldovich showed that if one associates an energy density with the cosmological constant, then this energy has the same effects as some bizarre form of energy that is associated with empty space (see below). This trend continued with other sporadic theoretical works until 1995, when cosmologists Lawrence Krauss of Case Western Reserve University and Mike Turner of the University of Chicago wrote a paper entitled "The Cosmological Constant Is Back." This paper was motivated by some observations suggesting the embarrassing situation in which the universe would have been younger than its oldest stars. The point is simple: If the universe was neither decelerating nor accelerating, but rather expanding at a constant speed, we could determine easily the time since the big bang by simply dividing the present distance to *any* remote galaxy by its present recession speed. Hubble's law, which states that the speed of the cosmic expansion is proportional to the distance, would ensure that we would obtain the same result irrespective of which galaxy we choose. If, on the other hand, the universe was decelerating significantly (e.g., with an omega of one), then the cosmic expansion would be *faster* in the past and the average speed would be higher (in fact, by 50 percent). This means that the universe would be *younger* (a shorter time since the big bang), possibly even embarrassingly younger when compared to the determined ages of stars. In particular, some clusters of stars known as globular clusters appeared around the mid 1990s to be older than the age determined for the universe, assuming that the latter was decelerating. If, on the other hand, the universe was *accelerating* (which a repulsive cosmological constant could do), the expansion would be *slower* in the past, rendering the universe with an *older* age. Therefore, Krauss and Turner reintroduced the cosmological constant to make the universe older, and thus to avoid the problem of a "mother" younger than her "children" (a universe younger than the stars within it).

The question of the age of the universe popped up under extremely unexpected circumstances in the following funny incident. In 1996 I organized a meeting at the Space Telescope Science Institute to discuss the age of the universe. One of the invited speakers was Nial Tanvir, a young astronomer from Cambridge

University. He started his talk by describing what happened to him at Heathrow Airport, on his way to the meeting. He had been stopped by a male security guard with long dark hair and wearing dark glasses. (Nial made a drawing of him.) The following conversation had developed:

Security guard: "Where are you going?"

Nial Tanvir: "To a scientific meeting in Baltimore."

Security guard: "What is the meeting about?"

Nial Tanvir: "What does it matter?"

Security guard *(insisting):* "What is the meeting about?"

Nial Tanvir *(with some impatience):* "We will determine the age of the universe."

Security guard *(with some surprise):* "And you think that you will be successful at that?"

Nial Tanvir *(sarcastically):* "We may come closer than you think."

Security guard: "But isn't there a problem with the ages of globular clusters?!"

In its reincarnation, the cosmological constant appeared in a different form—as a vacuum energy. The understanding of this concept requires some delving into the fascinating world of quantum mechanics. In quantum mechanics, even the vacuum, nominally empty space, is very lively. Far from being nothing, the quantum vacuum is full of continuously appearing and disappearing virtual particles or fields. As I noted before, quantum mechanics taught us that even if we perform an experiment and measure all possible quantities at one time, we can only predict *probabilities* for different results at subsequent times. Furthermore, Werner Heisenberg's *uncertainty principle* states that it is absolutely impossible to measure simultaneously and precisely both the position and the momentum of a particle. Any attempt to increase the precision in the measurement of one quantity will result in a deterioration in the measurement of the other quantity. Similarly, quantum mechanics allows energy not to be conserved precisely, for very brief periods of time. Thus, the virtual particles that roam through the vacuum owe their fleeting existence to the probabilistic nature of quantum mechanics. Namely, for brief periods of time, energy can be borrowed from the vacuum, for the appearance of particle-antiparticle pairs or their corresponding quantum mechanical waves. These pairs then annihilate themselves back into nothingness within about 10^{-21} seconds. At any given moment, the vacuum is bubbling with such virtual pairs. This richness of the vacuum is not merely a theoretical speculation; its reality has been verified experimentally.

For example, in 1947, Columbia University physicist Willis Lamb showed experimentally that two states of the hydrogen atom that were thought previously to have precisely the same energy really differed by a very tiny amount. This effect, known as the *Lamb shift,* was eventually explained by the realization that the shift in the energy involved a contribution from virtual photons and electron-positron pairs.

Another experiment that provides corroborating evidence for the existence of these virtual particles involves the *Casimir effect.* In the experiment, two parallel metal plates are held at a very small distance from each other (about ten times the diameter of an atom). In the space outside the plates, virtual pairs (which are represented by quantum mechanical waves) of many wavelengths can exist. Between the plates, on the other hand, only virtual pairs that can fit a precise number of wavelengths in the spacing can exist. Consequently, there are more virtual particles (waves) outside the plates than between them. This imbalance pushes the plates together with a measurable force, equivalent to about one ten-thousandth of the atmospheric pressure.

As perplexing as this quantum mechanical view of the vacuum may sound, the picture is somewhat further complicated by the fact that quantum mechanics distinguishes between a true vacuum and a false vacuum. The true vacuum is defined as the state of *lowest possible energy density.* In this state, the virtual fluctuating fields (which are represented by waves) can have all possible wavelengths, and they can travel in all possible directions. Consequently, when averaged over long periods of time, these fields tend to cancel out, in the same way that in the middle of Grand Central Station in New York you will not find that people move in any particular direction on average. In the false vacuum, on the other hand, fields do not cancel out; hence, the false vacuum is at a higher energy state than the true vacuum. I will come back to the false vacuum in my description of the inflationary model.

Returning now to the cosmological constant, recall that its effect (when it was taken to be positive) was to produce a repulsive force at large distances. The existence of a cosmological constant can be interpreted as the effects of this energy of the vacuum. Namely, all of these vacuum fluctuating fields have an energy associated with them. As I will explain later, the vacuum has the peculiar property of having a *negative* pressure (unlike a normal gas, which has a positive pressure), and this in turn has some important consequences for the gravitational force. For example, in Newton's theory of gravitation, the attraction is proportional to the density of mass. If we use the fact that according to Einstein mass and energy are equivalent, we could say that gravity is proportional to the *energy density*. In Einstein's general relativity, however, the *pressure,* in addition to the energy density, also contributes to the gravitational force. Specifically, the force is proportional to the *sum* of the energy density and three times the pressure. Since the pressure is *negative,* the gravitational force that it produces is *repulsive.* The vacuum thus acts precisely like a cosmological constant; namely, as an *antigravity force.* This can have quite dramatic consequences. As the universe expands and its matter spreads thinner to lower and lower densities, the gravitational attraction drops off. The cosmic repulsion caused by the vacuum can eventually dominate and even cause the expansion to *accelerate,* instead of the normal deceleration that is expected due to gravity. It is thus extremely important to attempt to understand how large the contribution to omega of the energy density involved in these quantum fluctuations can be. Put differently, we need to know the value (and sign) of the cosmological constant Λ. Unfortunately, most theoretical attempts in this direction so far have failed miserably. The first estimates, which were based on a simple set of assumptions, gave an infinite value. This in itself is not very surprising, however. The history of quantum electrodynamics—the quantum theory describing electrons, positrons, and electromagnetism—had been dogged by infinities for two decades through the 1930s and 1940s. In that case, it was found that because the theory required summing up an infinite number of contributions from virtual photons of unlimited energy, the sum was always catastrophically large. Only after physicists learned to handle these infinities correctly, by a redefinition of the mass and the charge of the electron to agree with laboratory measurements, did they find out that in fact they all cancel out. Interestingly, in spite of quantum electrodynamics' impressive successes, one of its founders, the physicist Paul Dirac, was never quite convinced of its correctness. He regarded the mere *appearance* of infinities, even if they can be made to cancel out, as a flaw in the beauty of the theory.

The next estimate (around 1980) of the vacuum energy density still gave a gigantic number. The point is that one might expect the most natural value of this density to be of the order of the energy density of the universe at the very beginning, when the universe was 10^{-43} seconds old (the epoch called the Planck era). In this case quantum effects on scales smaller than the typical size of the universe at the time (10^{-33} centimeters) are ignored, and the governing theory is that of quantum gravity. The resulting contribution to the energy density is still about 10^{123}. This just cannot be right, because it would mean, for example, that like vampires, we would never be able to see ourselves in the mirror. For such a huge cosmological constant the space between us and the mirror would expand too fast for light to ever be able to reach from us to the mirror or from the mirror back to our eyes. In fact, we would hardly be able to see anything at all. Many different string theories and supersymmetry (SUSY) give a variety of values for the vacuum energy density. However, most of these are still incredibly large (e.g., 10^{55} from SUSY in 1984). The problem is that even before performing any detailed observations on the expansion of the universe, in an attempt to determine the value observationally, we know that these huge numbers are totally wrong. The fact that we do not observe any noticeable effects of such a cosmic repulsion on the orbits of planets or on motions of clouds within galaxies means that the contribution of this vacuum energy (or cosmological constant) must be somehow totally canceled out, or at least be very significantly suppressed.

Because of these problems, most cosmologists regarded the introduction of the vacuum energy, or the cosmological constant, as an uncalled-for complication. Consequently, most pre-1998 guesses as to the value that will eventually be found for the cosmological constant put it precisely at zero. This is actually an interesting psychological reaction. It is not that anybody *knew* that the value has to be zero, but the simple estimates gave such unreasonably large numbers that most physicists felt that the only acceptable value would be zero. In a scorecard composed in 1998 by the Princeton cosmologist Jim Peebles, he noted that cosmological models involving a cosmological constant fail the test of aesthetics; and the University of Chicago cosmologist Rocky Kolb called such models "unspeakably ugly." In chapter 6, I will discuss some very recent observations that shed new light on this question. Here I should remark, though, that if the cosmological constant is not *precisely zero,* it may become more difficult to predict the ultimate fate of the universe. For example, if omega is 0.2 and there is no cosmological constant, then the universe will surely expand forever because gravity is just not strong enough to stop the expansion. If, on the other hand, omega is 0.2 and lambda is constant and *negative* (which in principle it could be, corresponding to an attractive force), then the expansion will surely come to a halt and the universe will collapse in a big crunch. This will happen irrespective of how small the contribution of the cosmological constant is (as long as it is *negative;* for example, –0.001). The reason for this behavior is clear. As the universe expands and its matter spreads ever so thinly, the gravitational attraction drops off; that is, omega approaches zero. Unless the cosmological constant is precisely zero, its effects will always eventually take over, and if it is attractive (negative constant), it will cause the universe to collapse.

Before I end this section, let me note on the issue of blunders that Henry Wheeler Shaw (who wrote under the pseudonym of Josh Billings) thought that unlike humans, nature never makes blunders (as the title of this section indicates). The full sarcastic quote, taken from *Josh Billings, His Sayings,* reads: "Nature never makes any blunders; when she makes a fool, she means it."

Euclides

Newton, one of the greatest scientific minds of all times, formulated one of the most fundamental laws of nature, the law of universal gravitation. His law proved extremely successful in explaining the motions of the planets and of bodies on Earth. However, Newton's formulation did not give any clue as to how this gravitational force is transmitted across the vast distances of the universe. It took another giant to complete the work—Albert Einstein.

One of the key concepts in Einstein's *special theory of relativity* is that space and time are inextricably linked. The familiar three dimensions of space combine with the one dimension of time to form an inseparable entity called *space-time.* While it is somewhat difficult for us to visualize a four-dimensional space, mathematically speaking one can construct spaces of any number of dimensions. Space-time is such a four-dimensional space, which uses as "directions" length, width, depth, and time. The only reason that this notion causes some discomfort is because we cannot actually *see* the time direction in the same way we see the other three dimensions. For example, when we look at an apple we see that it has a certain height, width, and depth. If we could see the time dimension, we would have at the same time seen the history of this apple from seed to being picked. Einstein's special relativity was formulated specifically to agree with Maxwell's theory of electricity and magnetism. However, Einstein found it more difficult to incorporate the effects of gravity. He spent the years from 1907 to 1915 in developing his theory of gravitation, which is regarded by many as the most beautiful physical theory to date. This theory, known as the *general theory of relativity,* was published in a series of papers in 1915–1916.

The central idea of general relativity is very *simple:* gravity is not some mysterious force that acts across space. Rather, the presence of mass warps space-time in its vicinity, in the same way that a heavy bowling ball resting on a thin rubber mat would cause it to sag and curve. If we rolled some small balls on this mat, we would see how their trajectories would be bent by the depression caused by the bowling ball. Similarly, Einstein argues, the planets move on curved orbits not because they feel the gravitational force of the sun, but because they follow the most natural orbits in the curved space-time produced by the sun.

To give another example of how a geometrical property of space could be perceived as a force, imagine two people who start to travel precisely northward from two different points on the earth's equator. Clearly they will meet at the North Pole. Now, if these people did not know that they were traveling on a spherically curved surface, they might have concluded that some force attracted them to one another, since they started traveling along parallel lines and yet arrived at the same point.

How did Einstein arrive at such a revolutionary idea as a *curved space-time* in lieu of a gravitational force? There were two major logical steps that he needed to take. The first, later dubbed by him "the happiest thought of my life," became known as the *equivalence principle.* Einstein realized that the fact that the force of gravity *is* proportional to the mass of the object on which it acts is reminiscent of something else. This something else is the inertial force. We are all familiar with inertial forces when we experience *acceleration.* For example, if we stand in a bus that suddenly starts accelerating, we feel a force pushing us backward. Similarly, when the bus decelerates, we feel a force pushing us forward (this is also why we buckle up with seat belts). The centrifugal force pushing us outward on a merry-go-round is also an inertial force. The inertial force *is also* proportional to the mass of the body on which it acts, and this, Einstein reflected, allowed for the possibility that gravity and inertial forces were really the same thing. Imagine that you stand in an elevator on a bathroom scale and someone cuts the cable holding the elevator. In this case, since both you and the elevator are free-falling with an acceleration of 32 feet per second per second (9.8 meters per second per second), the scale will record zero weight. Similarly, for astronauts in a space shuttle, the earth's gravity is precisely balanced by the centrifugal force and thus they are free-floating, or "weightless," feeling neither the gravitational nor the centrifugal forces. Einstein's remarkable genius was revealed in his conclusion from all of this—namely, that *the effects of gravity and of acceleration are exactly equivalent.* Thus, it is impossible to distinguish locally between a gravitational force and acceleration. Inside the falling elevator, it is impossible to tell whether you are weightless because the elevator is accelerating downward or because someone switched gravity off.

It took a second leap of imagination from the equivalence principle to the curvature of space-time. What does a curved space-time mean? It is easiest to answer this question if we examine the rules of geometry.

The geometry we are most accustomed to and that we learn in school was largely developed by the Greek mathematician Euclid (Euclides in Greek) in a thirteen-volume work called *Elements,* which appeared around 300 B.C. This geometry is thus known as *Euclidean geometry.* In this geometry, which was developed for a flat plane, the shortest distance between two points is a straight line, parallel lines never cross, and the sum of the angles in a triangle is equal to 180 degrees. There exist, however, other possible geometries, which are equally logical and which are based on similar sets of axioms—propositions that are assumed to be true. For example, on the curved surface of the earth, the shortest and the "closest to a straight line" path between two points is a segment of a *great circle.* This is a circle centered on the earth's center and connecting the two points. Airline flights from the United States to Europe follow great circles rather than following circles of latitude (which are not centered at the earth's center). The nineteenth-century German mathematician Georg Friedrich Riemann developed a geometry for spherical surfaces, which is known as *Riemannian geometry.* On a sphere, the shortest distance between two points is a segment of a great circle, lines that are parallel at one point (like two longitude lines at the equator) eventually cross (at the poles), and the sum of the angles in a triangle is more than 180 degrees.

There exists yet another geometry, known as *Lobachevskian geometry,* after the nineteenth-century Russian mathematician Nicolai Ivanovich Lobachevsky. This is also called saddle-shaped geometry, because it is applied on surfaces curved like a saddle. In this geometry, the shortest distance is a segment of a curve known as a hyperbola, lines that are parallel at one point eventually diverge from each other, and the sum of the angles in a triangle is less than 180 degrees. Interestingly, the Dutch graphic artist M. C. Escher created several works that depict excellent representations of Lobachevskian spaces. For example, in his series of woodcuts *Circle Limit I* to *Circle Limit IV,* he fills the entire area of a circle with identical shapes (fish, bats, or crosses). The shapes appear smaller and more crowded as the circular boundary is approached.

Einstein spent the years between 1907 and 1915 in search for a mathematical-logical framework in which to cast his ideas of gravity. Finally, it was through the equivalence principle that he found the answer. According to the equivalence principle, "weight" results either from a gravitational field or from acceleration. Similarly, "weightlessness" results either from free-falling in a gravitational field or from traveling at a constant speed along a straight line far from any fields. Einstein therefore reasoned that weightlessness is always associated with traveling along the path closest to a straight line in space-time. On the other hand, if you feel weight, he argued, then you are *not* on the straightest path. For example, since the earth is weightless in its motion around the sun (gravity exactly balanced by the centrifugal force), the shape of the orbit reveals to us the geometry of space-time. Thus, according to general relativity, the earth does not feel any force—it merely follows the straightest possible path through curved space-time.

To definitively implement these ideas, Einstein needed a mathematical theory of curved spaces. To his delight, he discovered that Riemann and Lobachevsky had developed precisely such theories. The theory of general relativity was thus born. Almost from its first appearance, the underlying simplicity and symmetry of the theory gained it many admirers among the greatest physicists of the time. The physicists Ernest Rutherford (who discovered the atomic nucleus) and Max Born (one of the first to understand the probabilistic nature of quantum mechanics) compared it to a work of art.

The next step was to apply general relativity to the entire universe. Here, due to the cosmological principle of homogeneity and isotropy on large scales, the curvature of the universe must be the same everywhere. As I noted earlier, Einstein originally thought that the universe has to be static, and this led to the introduction of the cosmological constant. The truly dynamic nature of the universe, however, was first revealed by the work of the Russian cosmologist Alexander Friedmann. In spite of a difficult life in Petrograd, Friedmann, a self-taught relativist, managed to show that the universe must expand or contract. Unfortunately, his findings were originally rejected by Einstein as being insignificant, and Friedmann's premature death in 1925 prevented him from seeing his model become the accepted cosmological model. The works of Friedmann, the Belgian Georges Lemaître, the American Howard

P. Robertson, and the Englishman Arthur G. Walker eventually demonstrated that there are only three possible geometries that can describe the universe as a whole. If the cosmological constant is zero, then these geometries correspond precisely to the possible values of omega.

First, the universe could be *closed.* This model corresponds to the case in which omega is larger than one. In this model the mass density is sufficiently high that gravity will stop the expansion and the universe will recollapse. Geometrically, this model corresponds to a space-time with a spherical or Riemannian geometry. Namely, the mass density causes the space to "curve back on itself," forming something like the two-dimensional surface of a spherical balloon. In such a universe, if you travel along a "straight line" (which would really be a great circle, as in Figure 7a), you would eventually return to your point of origin.

A second possibility is that the universe is *open.* This corresponds to a value of omega that is smaller than one. In this case, the gravitational field is too weak to stop the expansion and the universe will expand forever. In geometrical terms, while in a closed universe space curves back on itself to generate a finite volume, in an open universe space curves "away" from itself (as in Figure 7b), thus producing an *infinite* space. The geometry of an open universe can be approximately described by the surface of a saddle.

The third possibility is that of a universe in which omega is precisely equal to one—that is, the density is precisely equal to the critical density. As we have seen, this puts this universe on the borderline between eternal expansion and eventual collapse (it expands forever, but the expansion speed becomes closer and closer to zero). Geometrically, space in this case is *flat* and infinite (see Figure 7c). This geometry is identical to the Euclidian geometry we learn in school.

Thus, without a cosmological constant, *each type of geometry corresponds exactly to one prescribed fate of the universe.* If the cosmological constant, or the energy density of the vacuum, is not zero, we can

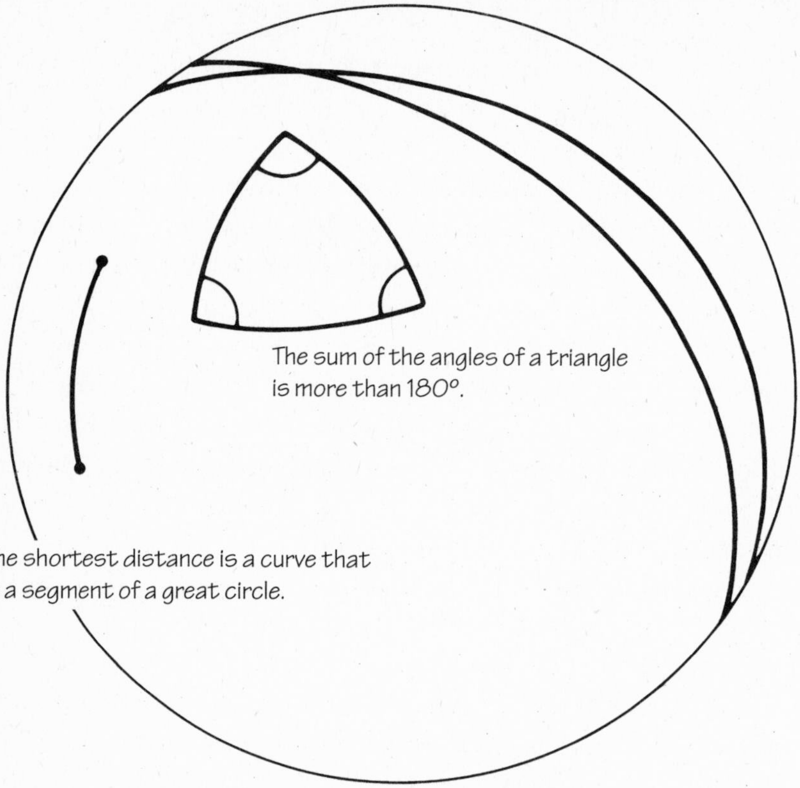

Figure 7a Closed Geometry

redefine omega. Recall that omega represented the fraction of the critical density that was contributed by all forms of *matter* (visible or dark). We can, however, ask what fraction of the critical density is contributed by truly *all* the forms of matter and energy, *including the energy of the vacuum.* I will call this new parameter *omega (total).* Clearly omega (total) is simply the sum of omega (matter) and omega (vacuum). Thus, an omega (total) equal to 1.0 corresponds to a flat geometry; an omega (total) less than one corresponds to an open geometry; and an omega (total) larger than one corresponds to a closed geometry.

As we shall see in the next chapter, there are good reasons to suspect that the geometry of our universe is in fact *flat.* This raises

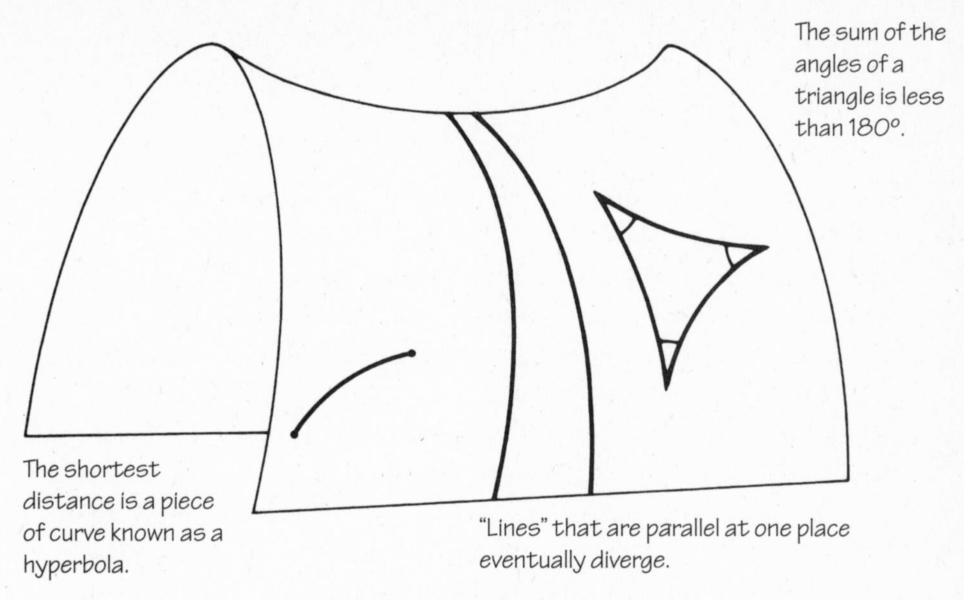

Figure 7b Open Geometry

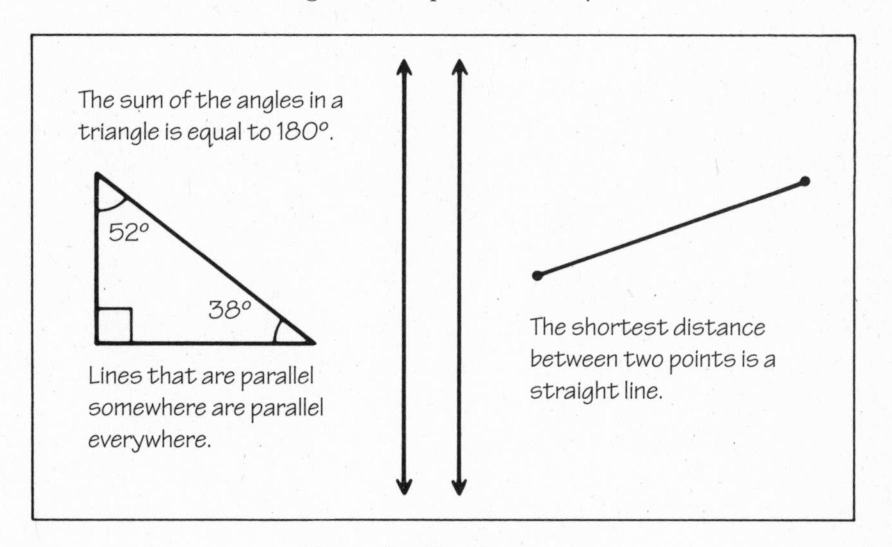

Figure 7c Flat Geometry

some interesting questions in relation to our everyday experiences. For example, the paper on which I am currently writing this section is flat. So is the desktop on which the paper rests and the floor on which the desk is standing. All around us we are surrounded by flat surfaces: the walls, the ceiling, the tabletops, the windows, and the doors. The geometry that we have all studied in school describes extremely well the flatness of the physical world in our immediate vicinity—in fact, so much so that we have almost come to believe that it is the only possible logical geometry. As we have seen, in reality, this is not the case.

But if our entire universe is truly flat, then a geometrical theory constructed by Euclid at 300 B.C., which reflects so accurately the structure of the limited space around us, also applies to the structure of the entire universe.

What perhaps is not fully appreciated is that in merely formulating his geometry, Euclid was in fact very much an experimental physicist, not just a pure mathematician. Einstein's general relativity has taught us that had the gravity of the earth been much stronger, we would have seen the space around us as being curved and not flat. Light rays, for example, would travel along curved paths rather than along straight lines. Euclid's geometry therefore reflects his observations (in the form of simple everyday life experiences) in the weak gravity of the earth. If the earth's gravity were significantly stronger, Euclid's geometry would have almost certainly been different.

6

When Inflation Is Good

In his great philosophical book *Moreh Nevukhim* (which means "guide for the perplexed" in Hebrew), Maimonides (or Moses ben Maimon), the most influential Jewish scholar of the Middle Ages, makes a parallel between humans and the universe. "Like the organs of the human body," he says, "all the particular things within the universe are intricately connected to form a united whole. Man is composed of body and mind, and all his modes of behavior are either forms of motion, which constitutes the action of his body, or forms of knowing, which constitutes the action of his mind." Maimonides then adds, "The universe constitutes of extension and thought, and all the modes of the behavior of things within the universe are either forms of motion, which constitutes the action of extension, or forms of intellect or understanding, which constitutes the action of thought."

Using this language, what I have described so far has more to do with the universe's "extension"; it is now time to turn to "thought."

The standard big bang model has three important consequences, or predictions, all of which have been spectacularly verified observationally. First, the model predicts that all the distant galaxies recede from one another, with a speed that is proportional to the distance between them. This is precisely what was observed by Edwin Hubble. Second, the model predicts that the entire universe basks in a highly uniform radiation bath, which today is but a cold reminder of the intensely hot distant past. The detection of this cosmic background radiation by Arno Penzias and Robert Wilson, and the determination of its properties by the COBE satellite, again confirmed unambiguously the big bang predictions.

Finally, the model predicts successfully the abundances of light elements, such as helium, deuterium, and lithium. These nuclei were synthesized from protons and neutrons during the first few minutes after the big bang.

All of these impressive successes pertain to the behavior of the universe from the time it was about one second old or older, till the present. However, about twenty years ago, astrophysicists realized that the standard big bang model also encounters some severe difficulties, especially at the universe's early infancy. Hence, these difficulties are associated with the very origin of the universe. The attempts to resolve these inconsistencies necessarily touch upon such questions as the origin of matter itself and of the cosmic expansion we observe. Here I will concentrate on three of these problems, since they are the most relevant ones for the discussion of beauty. These are known as the *horizon,* the *smoothness,* and the *flatness* problems.

Flatland's Expanding Horizons

The first problem is related to the fantastic broad-brush uniformity of the observable universe. In whichever direction we look, we observe the expansion to proceed in precisely the same way. This is truly amazing when one considers the fact that the universe may have started from "chaos"—a tempestuously irregular state. Recall that even the planetary nebulae, which are ejected from spherical and well-behaved stars, are almost never spherically symmetric.

Furthermore, as I described in chapter 3, variations in the temperature of the cosmic microwave background radiation from one direction to another amount to only one part in one hundred thousand!

Uniformity in any type of system requires first of all *communication* between the different parts of the system. For example, if one supermarket decides as a promotion to reduce its prices drastically, it takes some time for other supermarkets in the neighborhood to hear about it and follow, thus restoring uniformity in the prices.

Physical systems do tend naturally toward uniformity. This is one manifestation of what is known as the second law of thermodynamics. For example, the heat that flows from the vents of the heating system in a cold room will eventually spread and heat up the entire room evenly. However, the establishment of uniformity requires communication, and this takes time. Thus, the information that one corner of the room is colder than the other must first be "communicated" through the air molecules, and physical processes—in this case collisions between fast (hot) molecules and slow (cold) ones—must then operate to even out the temperature. If we were to measure the temperature at different points in the room before these processes had sufficient time to operate, we would not find the temperature to be uniform.

According to Einstein's special relativity, the absolutely fastest that any mass, energy, or information can travel is at the speed of light (even though it sometimes seems that rumors can travel faster). Thus, if we look at the evolution of the universe as a whole, at any given moment in the life of the universe there is a maximum distance that a light signal could have traveled during the time that has elapsed since the big bang. This is known as the *horizon distance.* We can expect to observe uniformity only on scales that are *smaller* than the horizon distance, since on larger scales communication does not have sufficient time to be established. Here, however, comes the problem. If the universe evolved according to the standard model, then it turns out that two points that are in opposite directions on the sky were *as far apart as ninety times the horizon distance* at the time when the background radiation was emitted. Yet we find the two points to have precisely the same temperature to within an accuracy of one part in one hundred thousand. How could two points that have not established contact "know" to equalize their temperatures to such a degree? To return for a moment to the example with the supermarkets, we would certainly be absolutely astonished if within one-tenth of a second of one supermarket lowering its prices, all the other supermarkets in town reduced their prices accordingly.

The physical reason for the existence of this communication problem is the slowing down of the cosmic expansion by gravity. The universe was expanding faster in the distant past. Thus, when the universe was, say, one thousand times smaller, it was actually *more than ten thousand times younger,* hence there was less time available for communication between the different parts. In other words, in the standard model the universe evolves too quickly, and there just isn't enough time for the uniformity to be established. This puzzle, of explaining how the universe can be so uniform over distances that are much larger than the horizon distance, is known as the *horizon problem.*

We should realize that the uniformity does not represent a real violation of the standard big bang model. If one *assumes* that the universe *started* from a uniform state, then it will remain uniform. This, however, is considered an extremely ugly solution to the problem, since it does not really offer any *explanation* as to why the universe is uniform. Rather, this solution simply relies on fine-tuning, by stating that the universe is uniform today because it was extremely uniform to begin with.

In fact, the problem is much worse. Even if one accepts the *assumption* of an initial uniformity on the largest scales, the standard big bang model requires yet another, even more extreme fine-tuning. When we examine the universe on smaller scales it appears anything but uniform, with galaxies and clusters of galaxies forming structures considerably denser than their surroundings. In order for galaxies and clusters of galaxies to form, some clumpiness or inhomogeneity on small scales must again be assumed. The early universe must have contained some small pockets of denser matter, which later grew to form the structures we observe today. If a small region of the universe is somewhat denser than its surroundings, then this region is braked slightly more by gravity. Consequently, its expansion lags behind that of its environment, and as a result of the reduced dilution it becomes even denser than the average. This effect snowballs as the expansion progresses, eventually producing individual galaxies. The same process of gravitational aggregation is responsible for the formation of clusters and superclusters. Regions that were originally slightly more crowded with galaxies will become even more packed as they lag behind the average expansion. The entire process is thus a perfect example of the rich getting richer. We may then ask how much clumpiness was required to start the process when the universe was, say, 10^{-43} seconds old (when gravity separated from the other forces). The answer is very unpleasant. The initial state must be *assumed* to be on one hand extraordinarily uniform, but on the other, not perfectly so. For example, the uniformity must exceed by far that expected from a normal gas in a state of thermal equilibrium. The inhomogeneities that are generated in such a gas by the mere fact that its molecules move randomly render it much too coarse for what is needed in the early universe. The standard model thus faces an additional ugliness in having to require an extremely fine-tuned initial state. This is sometimes referred to as the *smoothness problem.*

A third serious problem that the standard big bang model encounters is the *flatness problem.* In fact, I have already described the flatness problem in chapter 5. In the absence of a cosmological constant, if the energy density of matter in the universe is larger than some critical value—namely, if omega is larger than one—then the universe is closed. In this case space curves back on itself to form a finite volume with no boundary, like the surface of a sphere. In such a universe gravity eventually stops the expansion and the universe recollapses. In fact, in a universe in which omega is much larger than one, recollapse could occur so quickly that no observers would ever form to witness it.

If, on the other hand, the energy density is lower than the critical density (omega is smaller than one), then the universe is open. Space curves away from itself, approximately as in a saddle. In this case the universe continues in perpetual expansion.

If the energy density is precisely equal to the critical density (omega equals one), the universe is flat. In this case space is of infinite volume and described by Euclidean geometry. Such a universe expands forever, but with a speed that approaches zero in the distant future.

As we have seen in chapter 5, several methods for the determination of omega (of the matter, without the inclusion of the cosmological constant) give values in the range 0.3 to 0.4. But in order for omega to have such a value today, 14 billion years after the big bang, its value when the universe was one second old has to be *assumed* to be close to one to within a fantastic precision (to the fifteenth decimal place). Again, we see fine-tuning raising its ugly head. The failure of the standard model to provide any explanation of why omega began so close to one is known as the *flatness problem.*

We therefore see that in spite of its remarkable successes, the standard big bang model also has some serious flaws. It is important to realize, however, that none of the shortcomings described above is a genuine *inconsistency* of the standard model. With the appropriate assumptions about the finely tuned initial conditions, the horizon, smoothness, and flatness problems can all be solved without violating any laws of physics. However, *such a solution would be extremely ugly,* and it is therefore unacceptable on aesthetic grounds. As I explained before, a truly *beautiful* model of the universe should explain why the required conditions are essentially *inevitable.* As the Renaissance architect Leon Battista Alberti put it:

"Beauty is the harmony and concord of all parts, achieved in such a manner that nothing could be added or taken away or altered except for the worse." At the same time, any new model should also preserve all the striking successes of the standard model: the expansion, the cosmic background radiation, and the abundances of the light elements. This is another example of the natural selection process that guides physical theories toward beauty. A theory like the standard big bang model can progress in a certain direction as long as it is consistent with *all* the available information and it *passes* the aesthetic test. Once an obstacle is encountered on either front, a certain reevaluation must take place. Sometimes this soul-searching results in small modifications, sometimes in major ones, and occasionally it leads in an entirely new direction. Every once in a while, some brilliantly simple new idea arises, which leaves you wondering: "Why didn't I think of that?"

In 1981, the particle physicist Alan Guth, currently at MIT, proposed a major modification to the standard big bang model, with precisely the required exceptional qualities. His model, which expanded on previous ideas (mostly unknown to Guth) of Alexei Starobinsky in the Soviet Union, Katsuoko Sato in Japan, and others, became known as the *inflationary universe model.*

A Short Period of Inflation

The title of this section might have been taken from a speech by Alan Greenspan, the chairman of the Federal Reserve Board, but instead it describes one of the most extraordinary events in the history of the universe.

According to the inflationary model, the universe experienced an extremely brief period of incredibly rapid expansion, or *inflation,* resulting in an increase in size by a factor of maybe 10^{50}. All of this, however, was over by the time the universe was 10^{-32}

seconds old, from which point on the inflationary model agrees precisely with the standard model, thus preserving all of the latter's successes.

One of the attractive features of the inflationary model is that it adds a new element to cosmology, one inspired by particle physics. This is thus a perfect example of unifying links between the subatomic world and the universe as a whole. In quantum physics, elementary particles such as the electron are represented by fields, which are similar in nature to the familiar electromagnetic or gravitational fields. As we know, the latter are responsible for the action of forces—the electromagnetic and gravitational forces. The inflationary model adds a new such field, known as the *inflaton field.* As we shall soon see, this field is also responsible for a force. In fact, it imparts the entire universe with a repulsive force that is capable of stretching space. The physics that is involved is not of the type one studies in high school, and it is at present still somewhat speculative; however, it provides a superb example of the scientific endeavors to understand the origin of the universe.

In order to explain this phenomenon, let me start by using an analogy involving a surface resembling a hat. Imagine that we have a surface like the one shown in Figure 8, with a ball that is free to roll on this surface. Imagine in addition that the "hat" is connected to an oscillator that is able to shake it at variable strengths. Clearly, if the hat is vibrated very violently, then the ball is thrashed about so forcefully in every possible direction that its motion is hardly affected by the existence of the central crown. In such a case, if we were to take the average of the ball's position over a long time, it would be at the center of the crown. If, on the other hand, the hat was not vibrating and we just threw the ball randomly into the hat, most probably it would have settled in the low ditch on the brim surrounding the central crown. Finally, if the ball was in the central dent (as in Figure 8) and the vibrations were very weak, then the ball would remain in the dent, because it would not have sufficient energy to get over the barrier.

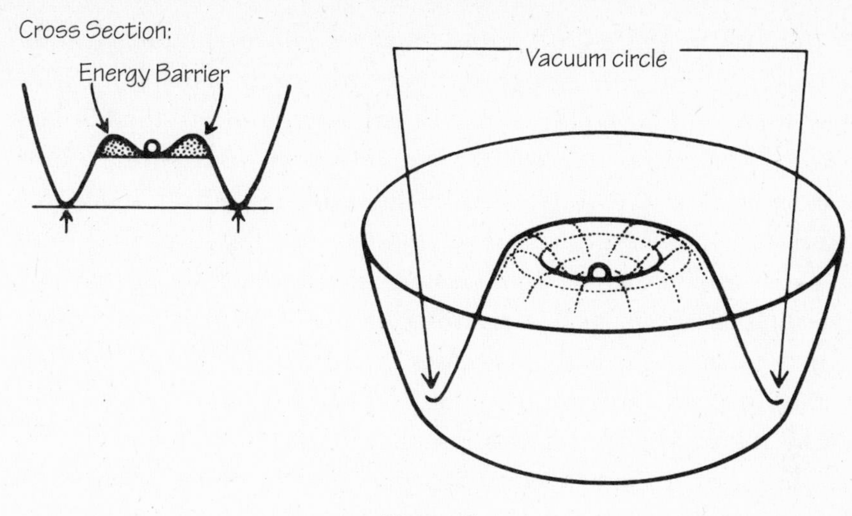

Figure 8

The vibrations of the hatlike surface are analogous to the temperature of the universe, with high temperatures corresponding to vigorous vibrations. The height of the ball above the surrounding ditch determines the values of the fields known as inflaton fields or their energy, in the same way that the height of a ball above the floor gives its gravitational potential energy with respect to the floor. The low ditch surrounding the central peak corresponds to the *lowest* possible energy state and thus to what is known as the physical *vacuum.* When the ball is in this vacuum circle, this corresponds to the temperature in the universe being (relatively) low. The inflationary model now envisages the following sequence of events. At times preceding 10^{-35} seconds, the universe is hotter than 10^{28} Kelvin. This is also the time of grand unification, when the strong and electroweak forces are unified. This high temperature corresponds to the hat vibrating violently. (In fact, at such high temperatures the hat resembles more a simple bowl, with no central crown.) As the universe expands and cools (the hat vibrates less and less violently), it eventually reaches a critical temperature (about 10^{28} Kelvin) for a phase transition (such as water turning into ice) to occur. From laboratory experiments it is well known, however, that if cooling is very rapid compared to the time it takes a normal liquid to complete the phase transition, then *supercooling* may occur. For example, water can be rapidly supercooled down to more than 20 degrees Celsius (36 degrees Fahrenheit) *below* its freezing point before it undergoes the phase transition that transforms it into ice. Similarly, in the production of glasses, the liquid glass is very rapidly supercooled to a temperature that is much lower than the freezing point. In the analogy with the hat, extreme supercooling corresponds to the ball being stuck in the central dent rather than falling into the surrounding ditch. That is, as the temperature drops the hat shape develops, and the vibrations cease very rapidly. By the time the ditch is fully developed, the vibrations are too weak to take the ball over the barrier, and it remains in the dent. This peculiar state, which is produced by an extreme supercooling, is known as the *false vacuum.* This name deserves some explanation. The normal, or "true," vacuum is defined as the state of lowest possible energy. In the hat analogy, this corresponds to the surrounding ditch. The central dent, while having a lower energy than its surroundings, is clearly not the state of lowest possible energy, since the ball could still roll to a lower energy yet—the true vacuum. However, in order to get out of the dent, the ball would have to penetrate a barrier (Figure 8), and this is not possible in classical physics if the ball does not have enough energy to jump over it. In quantum mechanics, however, there is always a probability for tunneling through the barrier. Because of its probabilistic nature, quantum mechanics allows even for processes that are energetically not allowed in classical physics to occur with a certain probability. Given enough time, such a tunneling or barrier penetration will always occur, in the same way that continuous shaking of an apple tree will eventually knock an apple loose. This phenomenon of quantum tunneling is very well understood and is used frequently in electronics. Thus, for this particular shape of hat, or energy distribution, the false vacuum always decays eventually to the true vacuum through the process of quantum tunneling.

Returning to the evolution of the universe, the scenario continues as follows. As the critical temperature for the phase transition is reached, supercooling occurs, and the fields are temporarily trapped in a false vacuum state (the ball is in the dent). Therefore, the phase transition (transition to the true vacuum) is momentarily delayed. However, through quantum tunneling in a small region of space, a bubble in which phase transition has occurred will be generated (like small patches of ice forming in freezing water, or water starting to boil in a few spots in a pot). Eventually, large parts of the false vacuum (although maybe not all) will be transformed to a true vacuum phase (equivalent to the ball falling into the ditch). Here, however, comes the most important point. Recall that the false vacuum, unlike a normal gas, has the peculiar property of having a large but *negative* pressure. This bizarre concept can be understood by the following thought experiment. Imagine that you have a bubble of *true vacuum* in a large box filled with *false vacuum.* The bubble will certainly grow, because the true vacuum is a lower energy state than the false vacuum, and systems tend to lower their energy. This means, however, that the pressure of the true vacuum is higher than that of the false vacuum, since expansion occurs from higher to lower pressure. Therefore, since the pressure of the true vacuum can be zero, the pressure of the false vacuum must be *negative.*

As in the case of the cosmological constant, this negative pressure has truly remarkable consequences for the gravitational force. As I explained before, whereas in the normal Newtonian theory the only source of gravity is the mass density, in general relativity the gravitational force is proportional to the sum of the energy (or mass) density and three times the pressure. Since in the false vacuum state the pressure is negative and it dominates the contribution of the energy density, the resulting gravitational force is *repulsive.* It literally acts like an antigravity.

Consequently, the expansion of the universe during that phase, known as the *inflationary era,* was accelerated. The universe effectively doubled its size every 10^{-35} seconds. This exponential expansion lasted only till the universe was about 10^{-32} seconds old, but during this tiny fraction of a second the universe increased its diameter by the gigantic factor of about 10^{50}. (If you start with the number 1, it only takes 167 doublings to get to 10^{50}.)

After this huge expansion, the phase transition has finally occurred. As in the case of the latent heat that is released when water freezes, the energy of the false vacuum was released in this phase transition. The latent heat reheated the entire expanded region to a temperature of almost 10^{28} Kelvin. From this moment (10^{-32} seconds) on, the inflationary model rejoins the standard model, and the universe continues to expand and cool just as described by the standard model. Since the reheating temperature is very near the critical temperature of the GUT phase transition, the smidgeon excess of *matter* over *antimatter* that is responsible for the complete dominance of matter in our universe is produced immediately after the inflationary era, just as described in chapter 3 (via *CP* violation).

I realize that in spite of my best efforts, the above description may not have been the easiest to follow (some readers may think that this is the understatement of the year); therefore, let me recapitulate. As the universe reached the critical temperature of 10^{28} Kelvin, at which the phase transition was supposed to occur, it supercooled. The universe was thus trapped in a state of false vacuum. The negative pressure of this false vacuum caused a stupendous expansion, inflating what was but a speck by a factor of 10^{50}. Parts of the false vacuum have eventually decayed through the process of quantum tunneling, reheating the entire inflated region. From there on the universe resumed its normal, more leisurely expansion, as in the standard model.

One more word about the gigantic inflation that a part of the universe experienced. It should come as no surprise that by doubling in size every 10^{-35} seconds or so, a huge expansion is obtained within 10^{-32} seconds. Imagine someone who decides to play roulette using a double-up system. He starts by placing $1 on red. If he loses, he places a $2 bet on red. If he loses again, he places a $4 bet on red. If he loses this, too, he places an $8 bet on red. The idea is the following: before placing the $8 bet, he lost $1 + $2 + $4 = $7. Therefore, if he now wins the $8 bet, he has a $1 overall profit (bets on color pay even money). Similarly, had he lost the $8 bet, too, but won the next one up, of $16, he would have still been ahead by $1 (since $1 + $2 + $4 + $8 = $15). So why aren't the casinos worried? After all, this person must win at some point. Well, the casinos are not worried at all, because after only nine losses in a row (which can easily happen) the tenth bet has to be $512, while most roulette tables only allow a maximum bet of $500! Therefore, successive doublings build up quickly to large numbers.

Answers in a Hat

The inflationary model solves the horizon and flatness problems easily, and in a most natural way. According to the inflationary model, our entire observable universe evolves from a region of space that is minuscule compared to the corresponding one in standard big bang. The tiny region to be inflated is much smaller than the distance over which communication has been established (the horizon distance) by the time of the start of inflation. Thus, this region has ample time to erase inhomogeneities and reach thermal equilibrium, unlike what is predicted in the standard model. It is this small and *homogeneous* region that is destined to become larger than our entire observable universe after inflation. It is therefore not surprising anymore that the COBE satellite measurements of the temperature of the cosmic background radiation reveal an astounding isotropy—the horizon problem is solved.

Inflation also explains naturally why our universe appears as flat as it does. In fact, inflation (in its simple form) has a rather well-defined prediction. It predicts that irrespective of what value it had before inflation, *omega (total) today should be almost exactly equal to one.* This prediction is easy to understand if we remember that an omega equal to one corresponds to a *geometrically flat* space-time. Inflation takes a minuscule region of space and inflates it 10^{50}-fold. If you take a spherical balloon and inflate it tremendously, its surface becomes (locally) flatter and flatter the more you inflate. For example, the surface of the earth appears to us to be flat locally. Alternatively, since our entire universe emerged from a speck, this is like examining a dot on a piece of paper; no matter how crumpled the paper is, its very tiniest regions are still flat. Recall that by omega (total) I mean the contribution from both all forms of matter and from the vacuum. An omega (total) equal to one thus corresponds to the situation in which the sum total of *all* the energy densities is equal to the critical density. If the cosmological constant is zero (no contribution from the vacuum), then the density of matter alone should be equal to the critical density.

Thus, inflation offers simple solutions to the horizon and flatness problems. In fact, as we shall soon see, it does much more. However, before describing inflation's other attributes, I should note that in spite of these impressive successes, soon after the inflationary model was proposed, it was realized that it also had one serious drawback. It was found that in the context of the original model, the phase transition itself (which marks the end of inflation) would form bubbles that create inhomogeneities that far exceed the observed ones (e.g., in the cosmic background radiation).

However, this problem (known as the *graceful exit problem*) has been largely solved, while retaining all of the model's successes, by some modifications to the original model. These modifications were introduced by the physicists Andrei Linde, then of the Lebedev Physical Institute in Moscow, and independently by Paul Steinhardt and Andreas Albrecht at the University of Pennsylvania. It is beyond the scope of the discussion in the present book to describe these modifications in detail, and I refer the interested reader to the excellent description in Alan Guth's book *The Infla-*

tionary Universe. Here I will only note that the key to the modified approach was to use a different shape for the surface on which the ball rolls (which determines the energy of the inflaton fields). That is, the hat in Figure 8 was replaced by a new hat, in the form of Figure 9. In this case, too, the evolution of the fields is similar to the motion of the ball over the surface. The new shape allows for a very gradual roll of the ball from the center of the plateau, followed by a slow roll into the ditch. In this case inflation continues while the fields are rolling, and thus the exit from inflation is less abrupt than in the original model.

In fact, in the years that have passed since the inflationary model was first proposed, it became clear that inflation really represents an entire class of models, rather than a single model. As many as fifty different forms of inflation have been discussed in the scientific literature since the idea's inception. While these various models differ in the details, and in the shapes of the hats they assume, the central idea remains the same.

At this point I cannot resist the temptation to digress a little, to describe another hat that is not really a hat.

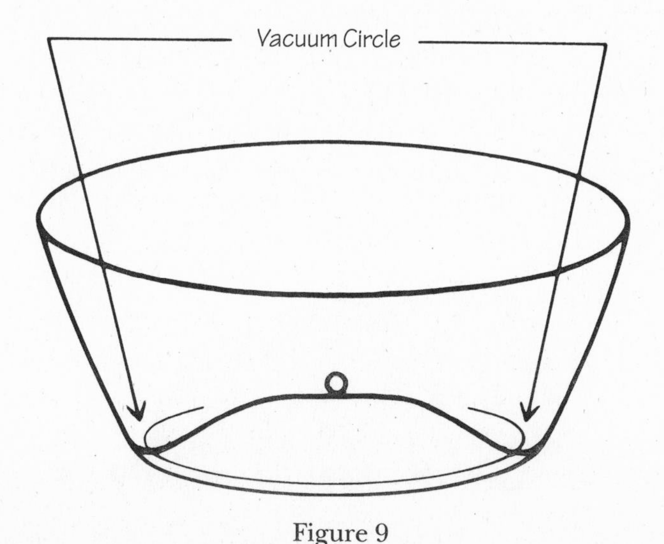

Figure 9

One of my favorite books of all time is *The Little Prince,* by Antoine de Saint-Exupéry. It describes how the author, while making a forced landing with his plane in the Sahara, meets the Little Prince, who arrived on Earth from a planet scarcely larger than a house. The book starts, however, with the following story, which the author tells about his childhood. When he was six years old, he saw a picture in a book of a boa constrictor in the act of swallowing an animal. The book said that boa constrictors swallow their prey whole, and then spend six months sleeping while digesting it. The six-year-old author then made his first drawing, which looked something like Figure 10. He showed this "masterpiece" to grown-ups and asked them whether the drawing frightened them. The grown-ups were surprised by this question and asked: "Why should anyone be frightened of a hat?" But, the author explained, this was not really a picture of a hat. It was a picture of a boa constrictor digesting an elephant. Since the grown-ups were not able to understand it, he had to draw a second picture, which now looked like Figure 11. To this, the grown-ups' response was to advise the author to abandon his drawings of boa constrictors, whether from the outside or the inside, and devote himself instead to geography, history, arithmetic, and grammar.

The author concludes this anecdote by saying that as an adult, whenever he met a person who seemed a bit more clear-sighted, he tried the experiment of showing him or her his first drawing, which he has always kept. If the person said, "That is a hat," then

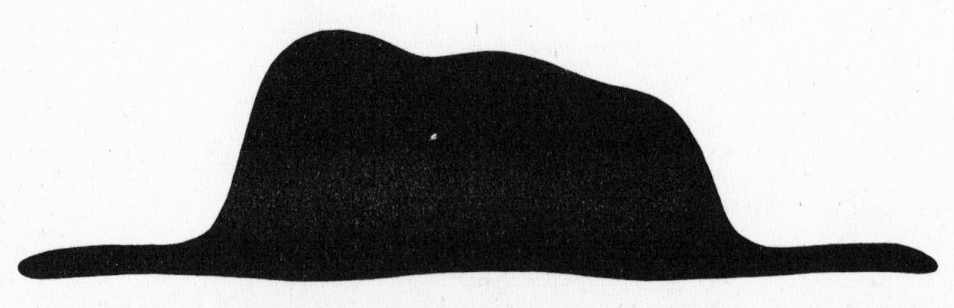

Figure 10

Figure 11

he would never talk to that person about boa constrictors, primeval forests, or stars. He would just talk about golf, bridge, and politics.

Cosmology Solved?

The inflationary model achieved much more than solving the horizon and flatness problems (which was an extraordinary achievement in itself). First, it may have also solved the smoothness problem—the generation of clumpiness, or density inhomogeneities, on small scales that can later develop into galaxies, clusters of galaxies, and superclusters. Inflation achieves this in a two-step process. First, inflation irons out any inhomogeneities that might have resulted from the initial conditions, in the same way that inflating a balloon smooths out all the initial wrinkles. Then, during the phase transition, tiny quantum fluctuations of the fields are generated, a bit like the bubbles in boiling water. These inhomogeneities appear however on quantum, or subatomic, scales, and would have been completely useless were it not for the process of inflation itself, which then inflates them to astronomically important scales. Inflation also predicts an extremely important property of these inhomogeneities: they should be *scale-invariant.* That is, the magnitude or strength of the clumpiness should be *the same on any length scale* of interest to astrophysics.

Inflation has not been successful thus far in explaining reliably the actual *magnitude* of the inhomogeneities, which was measured by the COBE satellite to be one part in one hundred thousand. This in itself may not be so surprising, since calculating the precise strength of the clumpiness requires a more complete understanding of the details of the grand unified theories than currently exists. However, the fact that the nonuniformities are scale-invariant was fully confirmed in 1992 by the Differential Microwave Radiometer (DMR) experiment (led by physicist George Smoot) on board the COBE satellite.

So, what has the inflationary model done for our understanding of cosmology?

1. Inflation explains why the universe is expanding in the first place; in the standard model this was simply assumed as an initial condition.

2. Inflation explains the origin of heat in the universe. The decay of the false vacuum that led to inflationary expansion and the release of all the energy when the phase transition took place is responsible for heating the universe and ultimately for the observed cosmic background radiation.

3. Inflation solved the horizon and flatness problems by suggesting that our entire *observable* universe originated from a minuscule region of the universe that inflated by an enormous factor.

4. Inflation provides an origin for the density inhomogeneities that served as seeds for the formation of all the structure that we observe in the universe. These inhomogeneities were created by the inflation of quantum fluctuations to astronomical scales.

This impressive list has led cosmologist Mike Turner of the University of Chicago to declare provocatively in 1998 that the answer to "cosmology solved?" may not be too far from "Yes!" This, however, raises the following important question: Is the inflationary model also *beautiful*? I should emphasize that by this I mean the *essence* of the model and not necessarily the details, which still need to be sorted out.

A Kandinsky Universe

I have identified beauty in a physical theory with symmetry, simplicity, and the Copernican principle. So one can put the inflationary model to the test and assess how it measures against these high goals.

Inflation passes the symmetry and simplicity test with flying colors. The inflationary model is in fact *based* on the relation between the microcosmos and the macrocosmos, on the application of the symmetry principles learned from particle physics to the entire universe.

Simplicity or reductionism is the inflationary model's hallmark. The central idea behind inflation is incredibly simple. By inflating a tiny part of the universe by a colossal factor, inflation solves immediately and simultaneously the horizon and flatness problems. The small scale of the region destined to be inflated ensures communication among its different parts, and results in a synchronized and fully homogenized behavior after the tremendous expansion. The expansion itself flattens out any curvature that might have existed, thus solving the flatness problem. As we have seen in the list in the previous section, in the true spirit of reductionism, inflation also solves at the same time a whole host of other puzzles, from the origin of universal heat to the provision of the seeds of structure. Furthermore, inflation has finally rid the model for the evolution of the universe of the dependence on the precise details of the initial conditions. For example, the theory no longer needs to assume *anything* about the initial homogeneity of the universe.

Since we have now established that the core of the inflationary model satisfies the first two requirements for a beautiful theory, it remains to be examined whether it is also consistent with the Copernican principle. In fact, inflation more than satisfies the Copernican principle; it gives the principle an even broader scope. According to the inflationary model, the unrestrained expansion that took place during the brief period of inflation took an infinitesimal speck and stretched it to a size vastly larger than the horizon of our present-day telescopes. *Our entire observable universe is but a tiny fraction of all that exists.*

Furthermore, if one consequence of inflation, known as *eternal inflation,* is correct (although at this point it is still quite speculative), then our entire universe may be just one in an infinite series of universes. This model was first suggested by Alexander Vilenken of Tufts University and later further elaborated upon by Andrei Linde, now at Stanford University.

As you recall, inflation was caused by the negative pressure of the false vacuum. At some point however, *a region* of the false vacuum decayed to the true vacuum (the fields rolled down from the central dent in the hat to the surrounding ditch), thus completing the phase transition and heating what was to become our universe to the point where it resumed its evolution according to the standard model. However, the regions of the false vacuum that did not decay continued during that time to inflate exponentially. In fact, calculations show that the rate of expansion is actually much *faster* than the rate of decay. Consequently, after a while, the volume of the regions remaining as a false vacuum became much larger even than the *entire initial volume.* Parts of these regions decayed to form other universes, while the remaining parts continued to expand. This process repeated itself ad infinitum, creating an infinite series of universes along the way. This series creates a *fractal* pattern; that is, if you could examine it on smaller and smaller scales with an ever-increasing resolving power, you would find that the same pattern of a universe surrounded by false vacuums is replicated again and again on all scales. Photographer Duane Michals has a similar sequence of photographs. In the first, you see a man looking at a photograph in a book. The second picture gives you a closer look at the book, and you see that the photograph in the book shows a man looking at a photograph in the book, and so on. Andrei Linde performed some computer simulations of this fractal sequence of universes with his son Dmitri, using different colors to represent different contributions to the energy of the universes. The resulting image resembled so much paintings by the Russian artist Wassily Kandinsky, one of the founding figures of abstract painting, that they called it a Kandinsky universe. Thus, if this picture of eternal inflation is correct, it implies the following: not only is our planet not at the center of our solar system, our solar system is not at the center of our galaxy; our galaxy is an ordinary galaxy; our universe does not have a center; and the universe is vastly larger than our observational horizon. It also implies that our universe may have been, for example, the number 10^{100} in line to be created, in an infinite series of universes!

Now, if this does not pass the test of the Copernican principle, then I don't know what does.

In the entire description of inflation and its consequences, I have deliberately not delved much into one important question. This question could, more than anything else, cast doubt on the beauty of our theoretical model of the universe.

But What about Omega?

Inflation (in its simple form) makes a very well-defined prediction about the value of omega. Since inflation produces a flat universe, it predicts that the omega (total) should be almost precisely equal to one. Namely, the *total* energy density (including that of the vacuum) should be almost precisely equal to the critical energy density. At the same time, most physicists are not thrilled (to say the least) by the prospect of having to rely on the energy of empty space, or the cosmological constant, to achieve flatness. Most physicists, if asked for their biased preference, would much rather have the omega of the matter be equal to one without a need for a poorly understood cosmological constant. The reluctance to accept a cosmological constant has also led to the construction of models known as *open inflation.* In these models, different shapes of the potential energy (the hat) result in inflation leaving the universe slightly curved (omega smaller than one), rather than precisely flat. As we shall soon see, however, recent observations appear to be in conflict with the possibility of a curved universe. I will therefore not discuss these models any further here.

Preferences aside, we have seen already that several methods for the determination of the contribution of matter to omega gave values in the range of 0.3 to 0.4, rather than 1.0. The question is then: are there any additional reliable methods one can use to determine both omega (matter) and omega (total)? Luckily, there do exist at least two such methods, one being extensively used already, and one that will be used in the very near future.

The first of these methods produced in 1998 a result that may prove to be the most exciting result in observational cosmology since the discovery of the microwave background.

The idea behind this measurement is very simple. If we will ignore for the moment the possibility of a cosmological constant, then it is clear that due to gravity, the cosmic expansion should be *decelerating.* This is similar to the fact that a ball thrown into the air is decelerating on its way up. What this means is that the universe expanded *faster* in the past than it does today. In fact, the larger the mass density in the universe, the larger the braking force of gravity and the greater the deceleration. By determining the rate at which the expansion is slowing down, we can therefore measure the mass density, or omega. How does one measure this deceleration? The idea is to measure the recession speeds of very distant objects. Since the radiation we see from such objects was emitted long ago, we actually measure the recession speed in the past. Such objects should therefore appear to be *receding faster* than expected from Hubble's simple proportionality between distance and speed.

The situation is slightly more complicated if the cosmological constant, or the energy density of the vacuum, is not zero. The cosmological constant (if positive) acts like a repulsive force at large distances. In this sense, the true vacuum, just like the false vacuum, has a negative pressure that produces a repulsive force. Consequently, if the vacuum energy density is not precisely zero, then the expansion of the universe could even be *accelerating* rather than decelerating. In such a case, distant objects would appear to be *receding slower* than the Hubble law would predict.

By measuring the *deviation* from the Hubble expansion law, astronomers can therefore determine the deceleration (or acceleration), and thereby the energy density, or omega. The pioneering work along these lines was done by Allan Sandage. There is probably no astronomer more deserving to be considered Edwin Hubble's successor than Sandage of the Carnegie Observatories. Sandage literally picked up where Hubble left off, in his relentless endeavors to determine cosmological parameters (and discovering a whole host of other phenomena along the way). Even today, at seventy-two, he continues to be one of the most active researchers in the field.

In order to be able to perform this measurement of deceleration or acceleration, however, two things are needed: (1) one needs to identify extremely bright objects, truly superluminous beacons that can be observed to distances corresponding to times when the universe was no more than half its present age (and preferably even younger); and (2) there must be an independent way to determine the distance to these objects, in order to allow measurements of the small deviation from the Hubble law. The best candidates for the use of this technique so far have proven to be *supernovae.*

Supernovae are among the most dramatic explosions in the universe. For a period of a few days, such an exploding star may outshine an entire galaxy. In our own Milky Way galaxy six supernovae have been recorded in the past millennium. Of these, the best known are one that was seen in A.D. 1054 and was well documented by the Chinese and native American "astronomers" of the time, and two that were seen within thirty-two years of each other and were observed by the famous astronomers Tycho Brahe (in

A.D. 1572) and Johannes Kepler (in A.D. 1604). Some astrophysicists remark sarcastically that the reason why no other supernova has been recorded in our galaxy in the past almost four hundred years is that there were no truly great astronomers since Tycho and Kepler.

Not all the supernovae are the same. Using spectra that give the light intensity at different wavelengths, and thereby serve as fingerprints identifying the presence of certain atoms, astronomers were able to establish the existence of two broad classes. Supernovae of one type (called, not very imaginatively, *Type I*) contain no hydrogen, while those of the other *(Type II)* are rich in hydrogen. The supernovae that are of interest to us here are a subclass of the supernovae Type I, known as Type Ia. These particular supernovae are used because they are the brightest and they allow for quite reliable distance determinations. Astrophysicists have developed fairly detailed models for the explosions that are responsible for Type Ia supernovae. These models involve compact objects known as *white dwarfs.* These are the remnant cores of very ordinary stars that were initially a few times more massive than our sun. As these stars evolve and approach the final stages of their lives, they eventually shed their outer layers (forming the planetary nebulae mentioned earlier), leaving a compact core behind. These white dwarfs are essentially the nuclear ashes of the dying star. They have a mass similar to that of our sun, but they are compressed by gravity to a radius like that of the earth. Consequently, they are typically a few million times more dense than water. There is a *maximum mass* that a white dwarf can have, which is about 40 percent more massive than our sun. This limiting mass is known as the *Chandrasekhar mass,* after the renowned astrophysicist Subrahmanyan Chandrasekhar, who first predicted that such a maximum mass should exist. Upon reaching the Chandrasekhar mass, white dwarfs that are made mainly of carbon and oxygen undergo a dramatic thermonuclear explosion that disrupts the entire star and sends its debris into space at speeds exceeding ten thousand miles per second. Many white dwarfs, like the one that will be the remnant of our own sun, simply fade away to end their lives with a whimper. Some, however, are born paired with a stellar companion. This allows them, under certain circumstances, to swallow material from the companion, thus driving themselves toward death with a bang.

Type Ia supernovae can be seen and identified even with present-day telescopes to distances that span about half of our universe's age, and with the *Next Generation Space Telescope* (NGST), to be launched in 2008, they will be seen to distances corresponding to less than one-twentieth the universe's lifetime (if they were occurring back then). Typically, the light from such an explosion takes about three weeks to reach its peak, and then it declines over a period of a few months. As I noted above, for individual objects to be useful cosmological tools, there must be a reliable way to determine their distance—they must be what are known as *standard candles* (namely, they all have the same brightness). This is a very simple concept. If we have a standard 100-watt bulb, and we observe it from different distances, we can use the observed light to determine the distance precisely, since we know that a source of light that is twice more distant appears four times dimmer. Similarly, if we know that objects belonging to a certain astronomical class all have precisely the same brightness (which is accurately known), then the distance can be determined from the observed light. Type Ia supernovae are not all exactly identical standard candles, but the small differences in their brilliance are relatively well calibrated. In particular, brighter supernovae are known to last somewhat longer than fainter ones, in a well-prescribed way. Therefore, careful monitoring of the duration of the bright phase allows astronomers to correct for the small differences. Type Ia supernovae are therefore superb distance indicators.

One Type Ia supernova explodes in a typical galaxy about once every three hundred years. Consequently, a monitoring program that follows, say, five thousand galaxies can expect to find in them one or two supernovae per month. The observing strategy is thus very simple: astronomers obtain images of the same piece of the sky a few weeks apart, and then search for points of light that have appeared or disappeared. Once candidate supernovae are identified, these become the targets for hectic observations with large telescopes such as the Keck Telescopes in Hawaii. These additional observations are the ones that identify whether the explosion truly represents a Type Ia supernova and determine the distance and redshift.

In recent years, the attempts to use Type Ia supernovae to determine omega were carried out independently by two teams. One, called the Supernova Cosmology Project, is led by astronomer Saul Perlmutter of Lawrence Berkeley National Laboratory in California, and the other, the High-*z* Supernova Team, (*z* is the letter that denotes redshift), is led by Brian Schmidt of Mount Stromlo and Siding Springs Observatories in Australia and Robert Kirshner of the Harvard-Smithsonian Center for Astrophysics. The observational campaign and full analysis of the results of a single supernova can last longer than a year, since the observers have to wait till the supernova basically fades from view. This is necessary in order to obtain a good image of the quiescent galaxy that hosted the supernova. The reason that the latter is so important is the following: the supernova appears as a bright dot that is superposed on the image of the galaxy. Therefore, in order to determine accurately the glow of the explosion itself, the background light of the host galaxy must be subtracted. The sharp images produced by the Hubble Space Telescope (HST), which are able to pinpoint the supernova location precisely, prove superior for this purpose.

Originally, the chances of success of these campaigns were considered rather slim by many astronomers, because of the smallness of the expected deviations (due to deceleration or acceleration) from the Hubble expansion law. However, the two groups, encouraged by early results, continued to move vigorously forward. I remember that about three years ago I received a phone call from Saul Perlmutter, who was interested in my assessment of their chances to be awarded observations with the HST. I assured him that I was convinced they would get the time because of the potential significance of the observations, and my conviction turned out to be correct.

At the time of this writing, each of the two teams observed and analyzed a few tens of supernovae, at a range of distances spanning a corresponding time interval of about one-half the age of the universe.

The results of these analyses came as a shock to the two teams, and resonated almost as a supernova explosion with the astronomical community. Distant objects appeared to recede *slower* than expected from the normal Hubble expansion law. Namely, *the results are consistent with an accelerating expansion of the universe.* Not only is our universe flying apart, but it does so at ever-greater speeds.

When I first saw these results—Robert Kirshner gave me a private preview prior to publication during one of his many visits to Space Telescope Science Institute—I must admit that I did not believe them. Since then, however, both groups have done a truly marvelous job of eliminating other possible interpretations. For example, since the main result is that distant supernovae appear dimmer than they should, one could think that cosmic dust obscuration could do the trick. However, just as they do in the earth's atmosphere, dust grains tend to filter out blue light more than red. Consequently, one would expect the supernovae to appear redder than expected (like sunsets), which they don't. Furthermore, in October 1998, the Supernova Cosmology Project team found one supernova (nicknamed for the Italian composer Tomaso Albinoni) that is more distant than all the others, but which is not as anomalously dim. This is contrary to the effect one would expect, for example, from dust obscuration, which should steadily increase with distance. Being less dim is easy to explain in an accelerating universe, however, since the acceleration was lower (and even became deceleration) in the more distant past, when gravity was stronger.

I personally have worked for many years on theoretical models for the progenitors of Type Ia supernovae, trying to identify from which types of stars the white dwarf slurps the material it needs for explosion. Prompted by these surprising observations, my colleague Lev Yungelson, from the Institute of Astronomy in Moscow, and I examined the possibility that distant supernovae are caused by different progenitors than nearby ones. While we showed that this was possible in principle, I also convinced myself that it was rather implausible. Finally, at the end of October 1998, a meeting of supernova experts and cosmologists was organized at the University of Chicago, for a comprehensive assessment of the two teams' results. At the end of three days of talks we all came out feeling that while surprises are definitely still possible, the results appear quite convincing. One possible alternative to accelerated expansion is that supernovae evolve during the cosmic history in such a way that the more distant ones (which occurred earlier in the universe's lifetime) are for some reason dimmer than the closer ones (which occurred relatively recently). This possibility needs to be further investigated by future observations.

The results of the two supernova projects (if fully confirmed) appear to rule out at a very high confidence level the possibility that omega of the matter is equal to 1.0, the aesthetically preferred result. Furthermore, the results are consistent with an omega of matter of about 0.3 to 0.4 and a vacuum energy or cosmological constant contribution to omega of about 0.6 to 0.7. Therefore, if these results are correct, then *the energy of our universe is dominated by empty space.* No wonder that *Science* magazine selected the accelerating universe as their major "Breakthrough of the Year."

The observant reader probably noticed that the above values are such that omega (total), which is the sum of omega (matter) and omega (vacuum), is consistent with the value of one, and thus with the prediction of the inflationary model.

Before I discuss the implications of this result further, however,

I would like to describe the second, and perhaps most promising, method for the determination of cosmological parameters.

The second method relies on *anisotropies* in the cosmic background radiation—namely, on differences in the temperature of the radiation from one direction to another.

Recall that the early universe, before neutral atoms formed, was opaque to radiation. Only when the universe was about three hundred thousand years old, and it cooled sufficiently for electrons to be captured inside atoms, did it become transparent. The cosmic background radiation, which started to roam all over space unhindered at that time, provides us therefore with a snapshot of the universe three hundred thousand years after the big bang. That was the time when the photons of the radiation experienced their last interaction with matter. These photons thus come from something like a surface that describes the conditions in the universe at that age. The situation is a bit like cars coming out of a toll plaza on a major highway. Just before the plaza cars are jam-packed and hardly able to move, but once they clear the surface determined by the toll booths they start to stream freely.

Variations (or anisotropy) in the temperature of the cosmic background radiation between directions separated by small angles across the sky correspond to clumpiness in the distribution of matter in the early universe. It is from these primeval density inhomogeneities that galaxies, clusters of galaxies, and superclusters eventually evolved.

Between the cosmic ages of about ten thousand and three hundred thousand years, clumps of the normal, baryonic matter could not start to grow. This is due to the fact that the matter was still opaque, and thus radiation was able to push on it (the fact that the matter is opaque means that it absorbs the radiation and feels its pressure). The radiation pressure therefore prevented matter from gravitationally collapsing to form clumps. No such inhibition existed, however, for the nonbaryonic dark matter, because the exotic particles of this matter are impervious to radiation. Consequently, the dark matter could start to clump and to form high-density regions long before the baryonic matter. Once the universe finally started to become transparent (at an age close to three hundred thousand years), radiation could no longer totally keep the baryonic matter from being drawn by gravity into the densest regions formed by the dark matter. In the early stages of this matter falling into these deep gravitational wells, however, radiation pressure was still trying (and partially succeeding) to push it back out. This resulted in something like an oscillatory, back-and-forth motion. You will observe a similar phenomenon if you try to push a floating wooden cylinder deeper into the water. The cylinder will oscillate up and down. These oscillations produced waves, just like the sound waves produced by the vibrating membrane of a drum, which undulated the surface from which the background radiation was emitted. Namely, they produced temperature fluctuations in the background radiation.

In the same way that the wavelengths of the waves created in the head of a drum depend on the size and geometry of the head, the longest wavelength that had time to oscillate in the cosmic background radiation depended sensitively on the geometry of the universe. As you recall, the latter is determined by the value of omega (total). In particular, calculations show that if the universe is flat—omega (total) equals one—then the differences in the temperature between two points in the sky are the *largest* when the points are about *one degree apart.* If, on the other hand, omega (total) is smaller than one (open universe), then the differences are largest for separations that are *smaller* than one degree. The reason for this qualitative dependence on omega is easy to understand. For a higher omega (denser universe, stronger gravity), the universal expansion *decelerates* more. Consequently, the distance to the surface at which the cosmic radiation was emitted is smaller, and thus clumps cover a wider angle on this surface. Similarly, a dime held closer to our eyes covers a much wider angle of our field of view than one held further away.

Experiments were therefore designed to measure the numbers of ripples of various sizes. A maximum in the abundance of ripples ("power") at a scale of about one degree on the sky would indicate a value of omega (total) of one, and a flat universe. Thus, if it is found that the largest differences in temperatures are between points on the sky that are about one degree apart, this will strongly suggest that our universe is flat, with an omega (total) of one. Data have been accumulating over the past few years from the COBE satellite, from balloons, and from ground-based observations. One set of measurements made from a telescope in Saskatoon, Saskatchewan, Canada, indicated a rise in the abundance of ripples toward a possible maximum at one degree. A second set, taken by the Cambridge Anisotropy Telescope in the United Kingdom, suggested that there are fewer ripples on scales smaller than one degree. Even more suggestive results came in mid-December 1998, from two experiments at the South Pole. The two microwave telescopes, named Viper and Python, take advantage of the very dry air at the South Pole to obtain a clearer view of the background radiation. Viper was looking for ripples on one side of the suspected peak, corresponding to angles smaller than one degree, while Python was surveying the other side of the peak, from one degree up to several degrees. The results, announced by Kimberly Cobel of the University of Chicago for Python and Jeff Peterson of Carnegie-Mellon University for Viper, seem to confirm the suspicion of a peak or a maximum in the abundance of ripples at around one degree. Python found the power rising up to one degree while Viper saw it declining on larger scales. Finally and most convincingly, measurements by the Microwave Anisotropy Telescope (MAT), located high on the slopes of Cerro Toco in Chile, showed a peak near one degree. These results were announced in September 1999 by the MAT team, a collaboration led by Lyman

Page of Princeton and Michael Devlin of the University of Pennsylvania. Thus, *there is a strong suggestion that omega (total) equals one.*

However, truly definitive values for both omega and the cosmological constant may come from two planned experiments. NASA's Microwave Anisotropy Probe (MAP, to be launched in 2000) and the European Space Agency's Planck, to be launched in 2007, will provide detailed full-sky maps of the cosmic background radiation, on scales as small as one-tenth of a degree. When the results from these experiments are combined with those from supernovae and other comprehensive surveys, the hope is that the cosmological parameters will be determined to within an accuracy of 1 percent.

At present, if one combines the results obtained from the supernovae during 1998 with those obtained from anisotropy measurements (during 1998 and 1999) one finds that the most favored values are *omega (matter) of about 0.3 to 0.4;* and *omega (vacuum) of about 0.6 to 0.7,* so that *omega (total) is consistent with being equal to one, for a flat universe.* The year 1998 was an extremely productive one for observational cosmology; it highlighted the transition to precision measurements in cosmology—a long way from merely observing that the night sky is dark!

7

Creation

As every stockbroker or meteorologist will tell you, it is infinitely easier to explain the past than to predict the future. When one deals with the entire universe, however, the validity of this statement becomes rather questionable. For example, we have seen that many lines of evidence indicate that the omega of the matter is about 0.3 to 0.4. This immediately suggests that even if the contribution of the cosmological constant is not known precisely, as long as it is not negative, the universe *is going to expand forever.* Such an eternal expansion can most probably lead to only one "sad" ending: a cold death!

What does this mean? What is actually going to happen?

The answer to this question is a direct consequence of the *attractive* nature of the gravitational force and the *repulsive* nature of the cosmological constant. Suppose for a moment that there was no cosmological constant. Gravity tends to pull different parts of any collection of masses together and make them contract. This is what causes stars to form from collapsing large gas clouds, galaxies to form from the merger and collapse of even larger clouds and collections of stars, and clusters to aggregate from assemblages of galaxies. All of these processes take time, first of all because of the enormity of astronomical distance scales, and second, because they are occasionally slowed down. Eventually such contractions, given enough time, lead to objects that are in equilibrium. Thus, stellar contraction is temporarily stopped when the star builds up a strong central pressure via nuclear reactions, which counterbalances the force of gravity. Similarly, galaxies attain stability by balancing the pull of their own weight with rotation (the centrifugal force) and the like. In a universe that expands forever (but with no cosmological constant), there is all the time in the world for many of these contractions to take place and equilibrium states to be obtained. Consequently, galaxies in a given group or cluster will tend to merge, to form one gigantic unit. For example, our own Milky Way galaxy will merge with the Andromeda galaxy in about 5 billion years. Concomitantly with all the mergers, since the universe continues to expand, it becomes more and more dilute, as the resulting clusters or superclusters increase their separations. Hence, in the future, the skies will become darker and darker. The individual galaxies will eventually run out of fuel (hydrogen) for forming new stars and the old ones will extinguish and die. As a result, even these aging galaxies themselves will grow dimmer and dimmer.

The situation is even gloomier if the value determined for the cosmological constant is correct. If the vacuum contribution to omega is indeed 0.6 to 0.7, as the most recent observations seem to indicate, the repulsion that this vacuum exerts will produce expansion at an ever-increasing rate. As gravity's resistance will grow weaker and weaker, the clusters that managed to form before repulsion took over will separate with increasing speeds, but no new ones will form. Generations of observers located in a galaxy in any of these clusters will probably watch with sadness as all the other clusters disappear from view. In fact, when the universe is about two hundred times its present age, any still-existing intelligent life will no longer be able to acquire any new observational data even on scales that we are presently able to observe.

If grand unified theories (GUTs) are correct, then protons will eventually decay. According to existing experiments, protons live longer than 10^{32} years. While this is a long time by all counts, in a universe with accelerated expansion, time is not an issue. Thus, all the atoms in dead stars, extinguished stellar remnants, and diffuse intergalactic gas will decay to such particles as electrons, positrons, and neutrinos.

Even black holes are not immortal. While in classical Einsteinian general relativity black holes are truly "black"—namely, no mass or radiation can escape them—when quantum effects are considered, black holes can actually emit radiation via a process known as *Hawking radiation.* This process was discovered by the famous Cambridge physicist Stephen Hawking in 1974–1975. The discovery followed a suggestion by the Israeli astrophysicist Jacob Bekenstein that black holes have a "temperature" associated with them, which is related to the strength of gravity at the *event horizon* separating the no-escape zone of the hole from the outside world. Hawking radiation is another wonderful example of reductionism or simplicity at work. Recall that the second law of thermodynamics states that the entropy, the amount of disorder of a system, always increases. Similarly, Hawking was able to show that when black holes interact with their environment, either by accreting gas from a nearby star or by colliding with another black hole, the area of their event horizon always increases. Bekenstein put two and two together and reasoned as follows: if black holes do not have any entropy (disorder) associated with them, we could violate the second law of thermodynamics by simply dumping all of our "mess" into black holes, thus decreasing the entropy (increasing the order) of the universe. If, on the other hand, we *identify* the area of the event horizon with the entropy of the hole, then the second law continues to hold, since while we decrease the entropy of the outside world, the increase in the entropy of the black hole itself compensates for this decrease. This concept was difficult to swallow at first, particularly because once you endow black holes with such thermodynamic properties as entropy, you are forced to conclude that they also possess a temperature, which is determined by the strength of gravity at the event horizon. However, any object with a nonzero temperature emits radiation, while black holes were thought to be absolutely black.

Here, however, Hawking was able to show that when quantum mechanics is brought into the picture, this is not necessarily the case. Hawking's proof was based on the richness of the quantum vacuum. Recall that the vacuum is really bubbling with activity, with pairs of virtual particles and antiparticles making a fleeting appearance before they are annihilated in a hurry. However, when all of this frenzy occurs near a black hole, something interesting may happen. Gravity can suck one member of a pair into the hole, leaving its partner not only unable to annihilate but in fact flying away from the hole. This process can therefore cause black holes to radiate, thus confirming the suggestion that black holes have an entropy, or disorder, associated with them. One of the most remarkable successes (although still with some limitations) of string theory in 1996 was an actual calculation (based on the microphysics) of the entropy of a black hole, by Andrew Strominger of the University of California at Santa Barbara and Cumrun Vafa of Harvard. Hawking radiation is particularly important for tiny black holes, while it is extremely slow for massive ones. The black holes that are the collapsed remnants of massive stars would survive for some 10^{66} years before evaporating completely. However, time is what an accelerating universe has in abundance. Hence, long enough into the future, even these seemingly indestructible objects will be reduced to nothing but radiation.

Is there anything that could give hope for a somewhat rosier future? Oddly enough, it is the existence of the cosmological constant that may save the day. In a paper published in February 1998, Jaume Garriga of the Autonomous University of Barcelona and Alexander Vilenkin of Tufts University offered a speculative scenario of a "recycling universe." In their picture, regions of true vacuum of our universe may "tunnel" back to the false vacuum, thus resuming inflation. The probability for such a process is tiny, since it is the equivalent of a ball rolling uphill to reach the top. However, in quantum mechanics even such a process is not totally impossible, and since the universe has an eternity for it to occur, it may. In such a case the universe may undergo an infinite succession of cycles from false vacuum to true vacuum and back. The amusing thing is that within each cycle, the entire history of a hot, symmetric universe may be repeated. Unfortunately, even if such recycling were to occur, there is no possibility for us to communicate our wisdom (for whatever it's worth) to future civilizations.

When I presented this generally bleak future in some popular lectures, there were always a few people in the audience who inquired: "So what was the purpose of this all?" Clearly, this question assumes that the universe has to have a purpose. This is in fact the basis of teleology—the theory that events can be explained only by consideration of the ends toward which they are directed. Teleological arguments have existed throughout the history of science. Aristotle, for example, believed that one cannot claim an understanding of any natural phenomenon unless one also understands its "final cause," or purpose. Teleological reasoning has made its way even into modern science, and it is, of course, discussed extensively by theologians. One manifestation of teleological overtones in modern cosmology is the anthropic principle, which I will discuss later. Here I will only note that "purpose" does not appear in the normal physics vocabulary, in which phenomena are explained in terms of physical laws. Physics has gone a long way in pushing a whole series of "whys" further and further back. Human curiosity, however, knows no bounds, and the realization that the future of the universe may hold nothing but a slow, cold death has only helped to further fuel the curiosity about how it all began. In describing current ideas of this very beginning, I will necessarily venture farther and farther away from direct experimental and observational physics, into the realm of pure theoretical speculations.

Neither Bears nor Woods

The eternal inflation model suggests that a region of false vacuum produces an infinite series of universes. In this eternally multiplying collection, universes may come in all sorts and shapes. While our universe is probably heading for a chilling end, others may recollapse to a big crunch. Furthermore, at the same time that universes die, new ones are continuously being born in their place. Therefore, from a *global* point of view, life could go on forever, albeit in different universes or parts (pockets) of universes. Thus, even though the lifetime of any individual civilization is finite, and its accumulated wisdom is doomed to be lost, civilizations can continue to exist at any time in some parts of the global universe (information cannot be transmitted, however, from one pocket universe to another). This seems to be an almost inevitable conclusion of many versions of inflation. Hence, the global future may not look so gloomy after all.

If this picture of eternal inflation is correct, then universes are a bit like human beings. Individual humans are born, live, and then die and disappear, but the totality of the human race continues. In some sense, therefore, this global universe is constant and unchanging.

The inflationary model was not the first to suggest a universe that exists in a *steady state*—namely, one that preserves the same appearance at all times. In 1948, the physicists Hermann Bondi, Thomas Gold, and Fred Hoyle suggested a model in which *our own universe is eternal and unchanging*. Since the universe was already known to be expanding, the steady state model suggested that matter is continuously being created, so as to keep the density constant. According to Bondi, Gold, and Hoyle, as galaxies are aging and flying apart, the gaps between them are filled with a new generation of young galaxies. Hoyle did not consider the fact that the creation of new matter contradicts the conservation laws of physics to be a serious objection, since, he argued, the creation of all the matter in the universe ex nihilo in one big bang was an even more objectionable idea. Furthermore, it was difficult to refute the creation of matter observationally, since the required creation rate was less than one atom in every billion years per cubic meter. However, the steady state model had a distinct prediction: extremely distant regions of our universe, which emitted their light long ago, should look on the average the same as nearby regions. The steady state model was debated for about fifteen years, but today it is considered to be completely excluded by observations. The fatal blow to the model was dealt by the discovery of the microwave background radiation, which confirmed a beginning from a primordial fireball. The latest beatings of an already dead horse came from two remarkable observations with the Hubble Space Telescope (HST). In 1995, the director of the Space Telescope Science Institute, Robert Williams, decided to devote much of his discretionary time on HST to observations of a single deep field—namely, to point the telescope to one fixed point in the sky for ten full days in order to obtain the most detailed image ever of the very distant universe. A similar observation was repeated with new instruments in 1998. The patches of the sky chosen to be observed were random, one near the Big Dipper in the northern sky and one in the southern sky, except that great care was taken to avoid regions that are obscured by dust in the Milky Way or that contain nearby galaxies or stars that are too bright. A quasar, a galaxy with an enormously active and luminous nucleus, was included in the southern field. The fields were observed through a series of filters in ultraviolet, blue, yellow, red, and infrared light, which allowed for a full reconstruction of color images. The results were absolutely breathtaking. Nearly three thousand galaxies were detected in the northern field and a similar number in the southern one, spanning distances that go back to when the universe was less than one-tenth its present age. I still remember the day when we saw the first images on the computer screen, just around Christmas of 1995. Anyone who thinks that scientists are reserved in their reactions should have seen this. People were literally jumping and shouting with excitement. Among the many scientific results that the two Hubble Deep Field observations produced and continue to produce, two are particularly relevant for my present discussion. When we look at galaxies nearby, most of the large ones appear in quite regular spiral or elliptical shapes. However, when we look at very distant galaxies in the Hubble Deep Field, many have highly irregular, disturbed shapes. A part of this effect results from redshift. When we look at a very distant galaxy, its light is significantly redshifted. Consequently, the light we see in the visible part of the spectrum was really emitted at shorter wavelengths, in the ultraviolet. This light is emitted by regions in which massive stars (which are intrinsically hotter) are still forming. We are thus seeing the knots in the galaxies that experience the most intensive star formation. However, the shapes of the galaxies are also different for other reasons. When the universe was one-tenth its present age, many of these galaxies had not yet settled into their final smooth shapes. Since the universe was also denser then, many of these galaxies experienced collisions with other galaxies, which are also partly responsible for the disturbed appearances. Furthermore, the faint galaxies also appear to be smaller in size than the galaxies we see nearby. These two effects have a simple interpretation in the context of hierarchical (or bottom-up) structure formation—the scenario in which structure is built from the smallest to the largest objects. We are literally seeing here galaxy building blocks—the pieces that eventually formed the galaxies we see today. The Hubble Deep Fields inspired a whole host of follow-up observations from ground-based telescopes and other satellites. In May of 1997, the Space Telescope Science Institute hosted a symposium devoted to results of all the observations resulting from the northern deep field. In the preface to the book containing the proceedings of this symposium I wrote, "Never in the history of astronomy has so much research effort been put into a completely blank piece of the sky!" The Hubble Deep Fields demonstrated beyond doubt that galaxies in the distant universe *do not* look the same as in nearby regions. If one still needed proof that our universe is not in a steady state, the two Hubble Deep Fields provided evidence in abundance.

Eternal inflation, however, does not predict that *our* universe is in a steady state. Rather, it predicts the existence of an infinite series of universes into the future. This raises a most intriguing question: if there is an infinite series of universes into the future, could there also be one into the past? Namely, is it possible that the *global* universe (which contains an infinite number of universes like ours) *did not have a beginning?* Clearly, the possibility of a universe that *always existed* and will *always exist* has implications that reach far beyond the scientific domain. In his book *A Brief History of Time,* cosmologist Stephen Hawking spelled it out rather bluntly: "So long as the universe had a beginning, we could suppose it had a creator. But if the universe is really completely self-contained, having no boundary or edge, it would have neither beginning nor end: it would simply be. What place, then, for a creator?" It is definitely beyond the scope of the present book to discuss this type of question, and in general, scientists' views on these topics tend to be rather naive, and are regarded as "curious and disappointing" (to use the words of Richard Elliott Friedman, author of *The Disappearance of God,* when commenting on Hawking's statement) by many theologians and philosophers. However, even for physics alone, disposing of the need for a beginning may be quite desirable. The point is that the laws of physics usually determine how a physical system will evolve from a *given initial state.* If the universe had a beginning, then in order to determine what that beginning was, we may have to use the laws differently, or use other types of laws altogether. For example, the law of gravity determines what trajectory an apple will follow once it detaches from the tree, but a different set of laws determines how the apple grew on the tree in the first place.

In order to rid the theory of this problem of initial conditions, Andrei Linde, one of the proponents of eternal inflation, speculated that this process indeed does not have a beginning. However, in 1994, Arvind Borde and Alexander Vilenkin of Tufts University proved (under a certain set of reasonable assumptions) that if the universe is not closed, then *it must have had a beginning,* even though it may always exist into the future.

It must be realized that what is called a "beginning" in the eternal inflation scenario is rather different from the concept that prevailed in preinflation physics, or in general in literary works, myths, and works of art, most of which were largely inspired by the biblical description. For example, in his 1925 painting *The Birth of the World* (currently in the Museum of Modern Art, New York City), the Spanish surrealist Joan Miró merged the process of the artistic creation of the painting with the subject matter of Genesis. Miró started this work by splashing onto the empty (void) canvas a chaos of stains and spills, mostly in gray, black, and brown. From these, he then created a few forms, like a large black triangle with a tail that may symbolize a bird. A bright red circle with a long yellow streamer is also very prominent in the picture and probably represents a comet. Finally, the picture contains a humanlike, white-headed figure, close to a black star. An important point to note is that the world to Miró, as in the Bible, means the earth and its surroundings, rather than the universe as a whole.

In the preinflationary-model era, the "beginning" usually referred to the single event we call the big bang that created *our* universe (or our "pocket universe," in Alan Guth's language). This, however, is not the concept of beginning that emerges from eternal inflation. The old concept would be equivalent to referring to the birth of a particular individual as the "beginning," while the new one, to the origin of mankind. Our pocket universe started with a big bang, but many universes may have existed before it. I noted before that physics always pushed "whys" and "hows" to earlier and earlier times: from the origin of the solar system, to the origin of the Milky Way, to the first three minutes in the life of our pocket universe, to the Planck era (when our universe was younger than 10^{-43} seconds). Now, however, eternal inflation (if correct) has pushed the beginning not only almost infinitely into the past, but also to some other arbitrarily remote pocket universe. This has very significant implications for eternal inflation as a *physical* theory. If eternal inflation really describes the evolution of the universe, then the beginning may be entirely inaccessible to observational tests. The point is that even the original inflationary model, with a single inflation event, already had the property of erasing evidence from the preinflation epoch. Eternal inflation appears to make any efforts to obtain information about the beginning, via *observations* in our own pocket universe (which, as I noted before, could be the 10^{100} in line to be formed), absolutely hopeless. The Viennese symbolist painter Solomon Moser created a painting entitled *Light,* in which a young man holding a torch is seen breaking through the clouds into the heavens. I find a great poetic resemblance between this painting, symbolizing the human capacity to attain "divine" knowledge, and the scientific endeavors to understand the origin of the universe.

Some readers may wonder about the origin of the title of this section ("Neither Bears nor Woods"). In the biblical book 2 Kings, in chapter 2, verses 23–24, the following description is given of a miracle performed by the prophet Elisha: "He went up from there to Bethel; and while he was going up on the way, some small boys came out of the city and jeered at him saying, 'Go away, baldhead! Go away, baldhead!' When he turned around and saw them, he cursed them in the name of the Lord. Then two she-bears came out of the woods and mauled forty-two of the boys."

The interpreters of this passage were divided in their opinions as to the nature of the miracle performed by Elisha. Some said that the woods were there, but originally were not inhabited by bears and Elisha made them appear miraculously. Others insisted that the woods were not there either, and Elisha made both the woods and the bears appear out of thin air. The phrase "neither bears nor woods" has come to express in modern Hebrew skepticism about the reality of something.

We saw that at least under some assumptions (that the universe is not closed) physicists managed to prove that the universe did have a beginning. The question is then: how did the universe come into being? Does physics even address such a question?

Nothing

Our pocket universe started with a big bang some 14 billion years ago, but that was not the beginning of the global universe, if we accept the (speculative) predictions of eternal inflation. Nevertheless, if the assumptions of Borde and Vilenkin are correct (universe not closed), then even the global universe may have had a beginning and did not always exist. While eternal inflation by its very nature prevents us from knowing *when* that beginning occurred, it does not prevent us from speculating about *how* it might have occurred. In the fifth century, Saint Augustine was warned by his colleagues that people who inquire about pre-creation are destined to inherit hell. However, as the story has it, he refused to believe that God would punish those with curious minds.

The first type of modern physical theories of creation assumed that space-time itself forms some sort of fixed *background* and that matter was created at some point in this background. A model along these lines was proposed by Edward Tryon of Columbia University in 1973. I should emphasize that in this model space and time are assumed to be eternally preexisting and it is only the materiate world that is created. What is matter created of? The vacuum! Recall that the quantum vacuum is bursting with activity. Virtual particle-antiparticle pairs pop in and out of existence for tiny fractions of a second. But, the reader may wonder, what about conservation of energy? The virtual particles, after all, take advantage of the probabilistic nature of quantum mechanics to allow them to exist for some 10^{-21} seconds, but even our own pocket universe exists for 14 billion years! Here Tryon involves a crucial role for gravity. Every piece of matter in the universe is attracted gravitationally by any other piece. The gravitational binding energy is always negative, because it takes work to separate two pieces of matter and bring them to a state of zero energy. In fact, in a closed universe, for example, one can show rigorously that the total negative gravitational energy precisely cancels out the positive rest-mass energy of all the matter in the universe. Thus, the total net energy of a closed universe is precisely zero. There is a suspicion that this may be true for the entire universe in general. Thus, conservation of energy is not a problem, because the zero-energy universe could emerge from a zero-energy vacuum.

At present, Tryon's model is not considered particularly attractive for several reasons. Space-time is a central element of general relativity. The force of gravity itself is merely a manifestation of the curvature of space-time. Thus, assuming an eternally preexistent space-time is not very satisfactory as a creation theory, since it leaves open the question of the creation of space-time itself. Furthermore, assuming the existence of a preexisting smooth flow of time, it is difficult to understand why creation would occur at a *particular* point in time. In fact, quantum mechanics would predict a certain constant probability of creation within any time interval. This type of philosophical difficulty (clearly without the concepts from quantum mechanics) was recognized long ago by the philosopher Immanuel Kant in his *Critique of Pure Reason* and indeed even by Saint Augustine before him in his *Confessions*. Consequently, a more appealing model would be one in which *space-time itself* comes into existence in the creation process. Enter Alex Vilenkin at Tufts University. In 1982, Vilenkin proposed a model in which the universe, including space-time, was created *"by quantum tunneling from literally nothing"* (emphasis added).

Let me first say a few words about this "nothing." When I was a kid we used to have the following joke: "Do you know how radio transmission works? Imagine you have a very long dog. You pull its tail on one end, its head barks on the other. Now, radio works exactly in the same way, only without the dog!" "Nothing" here means truly nothing; this is neither the physical vacuum nor even space-time. Earlier I used the example of a two-dimensional universe that is the surface of a spherical balloon. In that example, painted dots on the surface represent the matter, while the expanding surface itself represents space-time. "Nothing" in this example would correspond to a totally deflated balloon whose surface shrank to *zero*. In the language of geometry, which is the one used in general relativity, in the same way that there is a flat or Euclidean geometry and a spherical or Riemannian geometry, there is also an empty geometry, a space-time that simply does not contain any points.

This is the geometry that would characterize "nothing." Quantum tunneling is the process by which systems can penetrate barriers that in classical physics are impenetrable. For example, if you have a ball at the bottom of a well, then in classical physics, unless the ball is given enough energy to reach above the top of the well, it cannot escape. In the subatomic world, where we have to use the rules of quantum mechanics, an equivalent situation would be to have a subatomic particle trapped inside an atomic nucleus (due to the attractive nuclear force), with insufficient energy to escape. However, quantum mechanics teaches us that there is a nonzero probability for such a particle to escape and find itself outside the nucleus after all. Similarly, the nuclear reactions at the center of our sun require protons to approach each other to distances of the order of 10^{-13} centimeters (the range of the nuclear forces). Yet, if it were not for quantum mechanics, protons with energies corresponding to the sun's central temperature could not get any closer to each other than 10^{-10} centimeters, because of their mutual electrical repulsion. Quantum mechanics allows for a certain probability for the protons to penetrate this repulsion barrier. Such barrier penetration processes, which would be totally forbidden in classical physics but are allowed (with a certain probability) by quantum mechanics, are known as *quantum tunneling*. The fact that such barrier penetrations can occur in the subatomic world is a consequence of the probabilistic nature of quantum mechanics. In quantum mechanics, particles like the electron are described by waves. The wave is stronger in places where the electron is more likely to be found, and weaker in places where the probability of finding the electron is low. In the same way that parts of a water wave are able to pass around obstacles, there is a nonzero probability to find the electron penetrating a barrier.

Vilenkin applied these concepts to the entire universe. If particles can tunnel, Vilenkin wondered, why not entire universes, including space-time? His suggestion was therefore that an *empty geometry*—corresponding to "literally nothing"—made a quantum tunneling transition to a *spherical geometry*—corresponding to a closed universe. Vilenkin himself recognized in his paper that "the concept of the universe being created from nothing is a crazy one."

The universe created by Vilenkin's process is very tiny, smaller than an atom. However, this problem is very easy to overcome by invoking inflation to expand the minuscule universe to an astronomically interesting size. Vilenkin's model is clearly highly speculative and at this point not really testable by experiments or observations, but it offers the distinct possibility that the universe appeared not just out of thin air, but literally out of nothing. Today, one can but wonder about the fact that this outstanding scientist was unable to get a job as a physicist in the former Soviet Union and had to work as a night watchman in a zoo!

Alternative (and equally speculative) creation models have been proposed; I will mention only two of these briefly. In 1983, James Hartle of the University of California at Santa Barbara and Stephen Hawking of Cambridge University attempted to use the rules of quantum mechanics to describe the universe as a whole. In the same way that subatomic particles are described in quantum mechanics by *wave functions,* the entities that give the probability of finding a particle in a certain state, Hartle and Hawking constructed (in principle) a wave function for the entire universe. The interesting thing about this approach is that Hartle and Hawking assumed that at times earlier than the Planck time (10^{-43} seconds), the usual distinction between past and future ("time's arrow") breaks down. Time is treated then in the same way as space. In particular, just as our universe has no edge in space, it also has no edge in time—that is, there is no "beginning"! The universe in this picture was never created, nor will it ever come to an end; it always existed and it will always exist. It should be realized that the Hartle-Hawking scenario is at this stage no more than an interesting *approach* to the problem, since an actual calculation of the "universal wave function" (which would give probabilities for all the possible events in the universe's life) not only requires a precise knowledge of all the laws of nature, but is also enormously difficult to carry out in practice.

Finally, in 1998, Richard Gott III and Li-Xin Li of Princeton University speculated that the universe can be trapped in some sort of an endless cyclic loop, a bit like a dragon biting on its tail. Their model can be likened to the following situation. Imagine that in one of the *Back to the Future* movies, Michael J. Fox travels back in time, only to become his own father. Clearly, in this case it would be impossible to trace his origins or family tree very far back. Gott and Li propose that our own pocket universe branched off from this cyclic global universe (like a French horn, which after completing a loop opens into a horn) and is now following its own evolution (toward a cold death).

I have given all of these examples first as an illustration of how physicists are attempting to use the basic laws of physics to struggle even with such questions as the ultimate miracle of creation. But I also had another purpose. These examples illustrate how different theoretical models for the same phenomenon can be, at the stage when they are neither guided nor constrained by observational or experimental tests. While astrophysicists agree on most aspects of the cosmological model from the time the universe was about one second old and on, there can be a wild diversity of speculative opinions on creation. The reason for this difference is very simple: ideas about the older, "mature" universe have been "naturally selected" via trial-and-error encounters with observational facts, while ideas about creation have so far avoided such battle testing.

This, however, highlights an intriguing point. If the inflationary model and, in particular, eternal inflation are correct, then the possibility exists that theoretical models of the global universe, or even just of the preinflation era, will *never* be testable by observations. I make here an important distinction between *a difficulty* in testing theories *in practice* and a true *impossibility in testing them even in principle.* There is no philosophical problem with tests being difficult in practice. This situation still characterizes many areas of the life sciences, our understanding of the operation of the human brain, and many dynamical systems, for example. However, models that *prevent* testing, even in principle, are at some level seriously lacking. I will call such prohibition in cosmological models *universal censorship* (not to be confused with the term "cosmic censorship," which has been used in relation to black holes). Universal censorship goes against the principles of the scientific method, and in particular it violates the basic concept that every scientific theory should be falsifiable. The question is, therefore, if eternal inflation is correct: is there some way to avoid universal censorship and to obtain information about our pocket universe's ancestry?

While thinking about this problem in mid 1998 I happened to hit upon the following fascinating story. A group of researchers led by Karl Skorecki of the Technion in Israel and Mark Thomas of University College, London, managed in 1997–1998 to trace the origin of Jewish priests to about three thousand years ago, before the Temple of Jerusalem was built!

According to biblical accounts of the Exodus from Egypt, Moses' brother Aaron was selected to be the first Jewish priest, or *cohen* (the Hebrew word for "priest"). This designation was also bestowed upon Aaron's sons. The tradition continued through the ages, with male *cohanim* (the plural of *cohen*) passing on the status to their sons. What Skorecki and his colleagues realized was that in the DNA, the Y chromosome passes solely from father to son, precisely like the status of a *cohen.* By studying certain markers on the Y chromosome of 306 Jewish men, including 106 *cohanim,* Skorecki, Thomas, and their collaborators were able to show that the *cohanim* indeed have some Y chromosome features that make them distinct from other Jews. Moreover, the researchers found suggestive evidence that the shared genetic material stems from an ancestor who lived between 2,100 to 3,250 years ago—a time frame coinciding with the historical accounts describing Aaron and his successors.

Earlier I used the analogy between the relation of pocket universes to the global universe and that of individual humans to humankind. Thinking along those lines, I therefore wondered, could there be some "universal DNA" that identifies universes uniquely and could allow (in principle) for a similar type of tracing as with the Jewish priests? Clearly, something similar to such a universal DNA could exist *in principle*—the set of laws of physics and the values of the universal constants that determine the strengths of the forces, the ratios of masses of particles, and so on. If indeed different universes have different values for the constants and/or different sets of laws, then these could serve for "DNA testing" of universes. As crazy as this idea of a universal DNA may sound, I soon found out that I was definitely not the first to think in those terms. In a paper I did not know about until 1998, theoretical physicist Lee Smolin of Pennsylvania State University had already proposed in 1992 that universes develop according to some rules of heredity and natural selection (the idea is nicely explained in Smolin's book *The Life of the Cosmos*). Smolin speculated that new universes and space-times are born within collapsing black holes. He therefore argued that only universes in which the laws of physics allow many stars to collapse and form black holes can produce many successors. Since, he conjectured, daughter universes possess sets of laws and universal constants that are only slightly different from those of their parents, prolific universe producers will "give birth" to other prolific producers, and vice versa. Consequently, according to Smolin's speculation, the population of universes down the road is dominated by universes whose laws of physics permit the largest number of black holes to form. This prediction can actually be tested (at least in principle) in our own pocket universe. If true, this would mean that the laws of physics and the values of the universal constants in *our* universe have been naturally selected so as to be optimal for black hole formation. Namely, any tinkering with the values of the constants in our universe should result in a *reduction* in the number of black holes being produced. On the face of it, this does not appear to be the case. A reduction in the efficiency of nuclear reactions, for example, would result in more stars being unable to resist gravity's final pull and becoming black holes. Irrespective, however, of whether the particular details of Smolin's speculation are correct, I find the idea of a natural selection for universes appealing, especially because of the shortcomings of universal censorship I mentioned above. Universal natural selection at least generates some form of continuity across generations of universes, which, if fully understood, gives perhaps a hope of overcoming universal censorship. I would like to remark, though, that we can expect a much better assessment of our ability to understand both the preinflationary era and the global universe once a fundamental theory emerges, perhaps via superstrings or M-theory (a unified framework of string theories).

Since this section discusses creation ex nihilo, it is virtually impossible for me to ignore the fact that this topic contains so many theological and mystical undercurrents. There have been many attempts to make various comparisons between cosmological models, in particular between the big bang and religious descriptions of the moment of creation. Most of these concentrate either on the biblical account or on Far Eastern religions. Examples include Gary Zukav's *The Dancing Wu Li Masters,* Fritjof Capra's *The Tao of Physics,* and Richard Elliott Friedman's *The Disappearance of God.* Not being a religious person myself, but having great respect for other people's beliefs, I have to say that in spite of the fact that many of these books (and in particular the ones just mentioned) provide for truly fascinating reading, I feel that they often suffer from the same kind of naïveté that characterizes physicists' treatment of theological issues. Detailed comparisons open themselves to unnecessary criticism and sometimes even ridicule from insensitive scientists. I personally have always regarded the description of creation in Genesis as a wonderful metaphor and a truly poetic account of what was surely in ancient times, and to a large extent is still today, an incomprehensible event. This is actually nicely expressed by one concept in the Cabala, a somewhat mystical medieval movement in Judaism. In the Cabala, the creation of the universe is so inseparable from the concept of the deity itself that the creation of the universe actually occurs *within* God. As a last comment on this issue, let me remark that even Nietzsche, quoted often to have said (in *The Gay Science*) that "God is dead. God remains dead. And we have killed him," says in another place (also in *The Gay Science*): "Even we seekers of knowledge today, we godless anti-metaphysicians still take our fire too, from the flame lit by a faith that is thousands of years old . . . , that God is the truth, and that truth is divine."

There is another aspect of cosmological models of creation that potentially touches upon the limitations of human understanding. Alex Vilenkin concludes his 1982 paper "Creation of Universes from Nothing" by stating: "The advantages of the scenario presented here are of aesthetic nature. . . . The structure and evolution of the universe(s) are totally determined by the laws of physics." This (if true) is indeed a remarkable achievement of the laws of physics. If, for example, the appearance of the universe is indeed the result of a quantum tunneling, then this would mean that creation was an inevitable outcome of the laws of physics. But that would leave open the question: what is the origin of the laws of physics themselves? As incredible as it may sound, a genuinely fundamental theory (like the string or the M-theory) could (in principle) perhaps provide an answer even to this question. Namely, the theory itself would provide only *one option* for what the laws could be, so that the theory would be *self-consistent* and at the same time explain the universe, from the subatomic world to the largest scales. One point that string theory has already demonstrated is that it absolutely has to be *a quantum theory,* with its inherent probabilistic nature. This is a remarkable discovery, since until now physicists have always been used to the concept of a classical limit—namely, that whenever one moves from the microscopic world of subatomic particles to the macroscopic world (of objects like tennis balls), one finds that the mathematical description of the phenomena changes smoothly from a quantum, probabilistic picture to a classical, fully deterministic picture (like Newton's laws of mechanics). In quantum mechanics, the behavior of, say, an electron is governed by a *wave function*. The evolution of this wave function is determined precisely by the theory, but the wave function in itself determines only the *probability* for certain future events, not the actual future itself. In classical mechanics, on the other hand, Newton's laws determine inescapably the future events in the "life" of a tennis ball. The transition from the quantum mechanical to the classical description occurs when the objects involved are large enough for their wave-probabilistic nature to become unnoticeable. String theory has shown that on small scales physics has to adopt the quantum description. Thus, one could say that the theory itself explains in some sense why there should be quantum mechanics. In this respect the theory determines its own laws.

Nevertheless, I would like to clarify that even though string (or other) theories may reduce the fundamental *theory* to one option only, this does not necessarily mean that there is only one possible *universe*. In the same way that one equation describing the hydrogen atom can still have many solutions, corresponding to the different energy states in which the atom can be, one equation describing the universe may still have many possible solutions, and therefore many potential universes.

I have now "traveled" both to the somewhat more predictable but gloomy distant future and to the more imaginative distant past of the universe. Returning to the observable present, an important question is: how does the discovery of accelerated expansion (if fully confirmed) affect the beauty of cosmological models?

The Disappearance of Beauty?

The most recent observations, from distant supernovae and from the anisotropy of the microwave background, indicate that our universe is *flat*—namely, omega (total) is equal to 1.0. The contribution of different components to the "filling of the bucket" of the energy density up to the critical value is roughly as follows. Stars and luminous matter do not contribute more than a meager 0.005 or so; neutrinos contribute an unknown fraction, surely more than 0.003, but no more than about 0.15; baryons contribute about 0.05. The bulk of the energy density comes from cold dark matter, about 0.3 to 0.4, and the energy density of the vacuum (the cosmological constant), about 0.6 to 0.7. We may ask: is all of this in conformity with our ideas of beauty?

First of all, the flatness in itself *is* beautiful, since it is fully consistent with the simple prediction of the inflationary model. As I noted before, the mere fact that to achieve flatness the universe has to rely on several sources of energy density, such as baryonic matter, neutrinos, and cold dark matter, is not in itself necessarily ugly, because there is nothing particularly fundamental about the different forms of matter. Ultimately, what elementary particles exist and what masses they have is, it is hoped, determined by the underlying fundamental theory, which in itself is governed by symmetry and simplicity. The situation becomes more questionable, however, when the cosmological constant is considered. It is certainly the case that about twenty years ago most, if not all, theorists would have subscribed to a zero value for the cosmological constant and an omega of one that relies only on contributions by matter. What exacerbates the situation is the fact that, as I explained before, the most natural value theoretically expected for the contribution of the vacuum energy to omega is about 10^{123}, while the measured value appears to be 0.6 to 0.7. Thus, apparently by some mysteriously precise process, the contribution of the virtual particles of the vacuum has been stripped from its most natural value to 123 decimal places, leaving only the 124th place intact. This does not appear at first sight particularly *simple,* thereby violating a basic requirement for beauty. As an aside, I should point out that in principle, the situation could even be much worse than we think. Within the errors that are still possible in the values of the different contributions to omega, it could be that those contributions add up (God forbid!) to an omega (total) of 0.9 or 1.1, rather than 1.0. If this were the case, then even the most basic prediction of inflation in its simplest form would be jeopardized. I will ignore here such horrifying possibilities, simply hoping that nature has some mercy on us in our attempts to understand it. In any case, the future anisotropy experiments—the Microwave Anisotropy Probe, or MAP, and the Planck probe—will, it is hoped, settle the question of the precise values of the cosmological parameters.

So, how should physicists feel about a *flat* universe that (surprisingly) is dominated by the energy density of the vacuum and of dark matter? The first important point to emphasize is that the fact that physicists and astronomers were surprised should not in itself be taken as evidence for ugliness. A limited understanding of what is truly fundamental can lead to gross misconceptions regarding the question of what constitutes a violation of aesthetic principles. A famous example is related to the discovery of the particle called a muon. In the *standard model* of elementary particles, which is based entirely on beautiful symmetry principles, the electron appears to have precisely the same status as two other particles—the muon and the tauon. Each one of these particles has a neutrino partner associated with it. Yet when the muon, which is 207 times more massive than the electron, was first discovered, the physicist Isidor Rabi asked with some indignation: "Who ordered that?" Thus, for physicists to be surprised usually simply means that the theory that existed at the time was lacking, rather than that aesthetic principles should be abandoned. A second example is related to the confusion between the symmetry of *shapes* and of *physical laws* I discussed in chapter 2. Starting with the ancient Greeks and for many subsequent centuries, the prevailing prejudice was that planetary orbits should be perfectly circular, because of the beauty associated with the symmetry of circles under rotation. This prejudice was so deeply rooted that, as the famous Cambridge astrophysicist Martin Rees describes in his book *Before the Beginning,* when Kepler discovered that the orbits are actually elliptical, even Galileo, who was a contemporary and correspondent of Kepler, was upset. Of course, elliptical orbits do not represent any violation of "beauty." The true implication of the symmetry of the *law* of universal gravitation under rotation is that elliptical orbits with *any orientation* are allowed. It is therefore not impossible that the prejudice against a dominant contribution by the cosmological constant merely reflects our present ignorance concerning where the ultimate theory is going to lead us. However, even if we accept this possibility, there is another aspect of the value of the cosmological constant (or density of the vacuum) that is quite bothersome.

As the universe expands, the average density of matter (of all forms) *decreases* continuously, from enormous values at the very early universe to the present value, which is 0.3 to 0.4 of the current critical density. For example, the matter in the universe was some 10^{27} times denser when the universe was one minute old than it is today. At the time when the first galaxies formed, the universe was still more than 100 times denser than at present. On the other hand, the density of the vacuum, as represented by the cosmological constant, is assumed to have stayed *constant* all this time (as its name implies) at its present (apparent) value of 0.6 to 0.7 of the current critical density. Thus, the contribution of the cosmological constant to omega was about 10^{27} times smaller than that of the matter in the one-minute-old universe. Matter still contributed 100 times more than the cosmological constant when galaxies appeared on the cosmic scene. It is only very recently that the cosmological constant got the upper hand over the plunging matter density, since their contributions are still very close to each other (0.6 to 0.7 for the cosmological constant compared to 0.3 to 0.4 for the matter). Almost throughout the entire universe's history, the density of matter was higher than that of the vacuum, and we happen to be living in the *first* and *only* time in cosmic history when the density of matter has dipped below that of the vacuum. On the face of it, this represents a serious violation of the generalized

Copernican principle (and is therefore ugly), because it implies that we live in a very special time.

What does all of this mean? Is it possible that we have come all this way, where in every step along the path our belief in the beauty of the universe has only been strengthened, to see it all collapse at the very end? I remember the first time that this realization hit me, in early 1998. I had a feeling in my stomach similar to the one I had in 1975, when I heard that somebody had carried a knife into the Rijksmuseum in Amsterdam and managed to gouge twelve deep slashes into Rembrandt's masterpiece *The Night Watch.*

The two major questions that the recent tentative findings on the cosmological constant pose are therefore: (1) why is it so small, but yet nonzero? and (2) why is it taking over the dominance of the cosmic density exactly now?

These are not easy questions, nor was their appearance particularly welcomed by most physicists. As I noted before, the cosmological constant has a rather inauspicious history, having been introduced by Einstein for the wrong reasons (to produce a static universe). Consequently, some physicists are hesitant even to grant this notorious "fudge factor" a physically fundamental legitimacy.

In an attempt to give a relatively quick answer to the above questions, some cosmologists have resorted to the *anthropic principle,* which invokes a role for intelligent life. Before I can examine this potential role in detail, however, I must address the question of how this intelligent life appeared in the universe in the first place. To this goal, there is no escape from taking a closer look at the one place where we know that life exists—the earth. Furthermore, in some sense, a crucial test for the Copernican principle would come from establishing whether we are alone in the cosmos or not.

8

The Meaning of Life

The emergence of life on a planet, and in particular intelligent life, surely requires a whole series of local conditions conspiring together to generate an optimal environment. However, even on the more global scale, if symmetry breaking were not to occur in our universe, there would be only one force and one particle—certainly no stars and no planets. Similarly, had *CP* violation not created a tiny excess of matter over antimatter in the primordial fireball, the universe would have been devoid of any matter today. The laws of physics in our universe, and the values of the various universal constants (such as the strengths of the forces) have allowed stars and planets to form and atoms to combine to form the complex phenomenon that we call "life." How did this all happen? And how "special" did our universe have to be for it to happen?

In 1970, the British mathematician John Conway invented a remarkable mathematical game called *Life.* Conway's goal was to generate a miniworld that is based entirely on mathematical logic, in the sense that a simple set of rules predetermines its evolution in every step, but that is still unpredictable. The game was introduced to a general readership in Martin Gardner's "Mathematical Games" column in *Scientific American* in October 1970 (and in Ivars Peterson's book, *The Mathematical Tourist*), and it proved to be habit forming for many. Since this game has a few interesting properties that are similar to the real evolution of life, I will describe the rules very briefly.

The game is played on an infinite (or very large in each dimension) checkerboard (or on the computer). Each "cell" has precisely eight neighbors (Figure 12a). If a cell has two neighboring cells that are alive (contain a black dot), nothing happens; the cell stays just as it was, alive if it was alive or empty if it was empty. If an empty cell has three living neighbors, this leads to a birth, which fills the empty cell in the next step. If the cell was already alive, it stays alive. If a living cell has four or more living neighbors, the overcrowding kills it (the cell is emptied in the next step). A living cell also dies (of isolation) if only one or none of its neighbors is alive. At every step in the game, each cell is checked and marked as giving birth, dying, or surviving, and all the changes occur in the next step.

Soon after the game was invented, the many *Life* aficionados discovered that it contains an endless number of possible patterns of evolution of this "life-form." For example, a given eternally unchanging state could have different predecessors (Figure 12b; try applying the rules to follow the evolution), or a very simple initial state could evolve into a pattern that oscillates back and forth between two

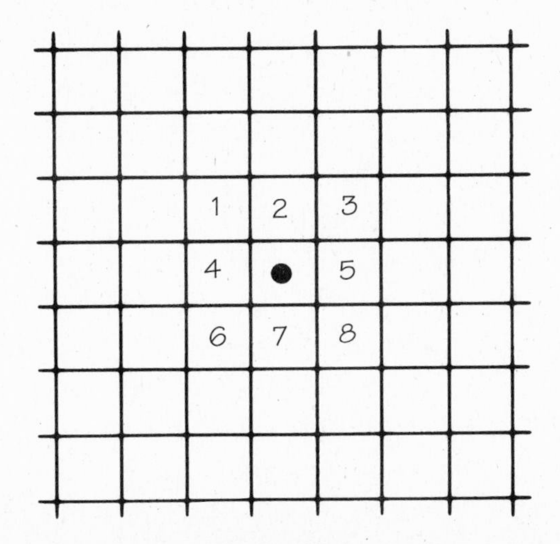

Figure 12a

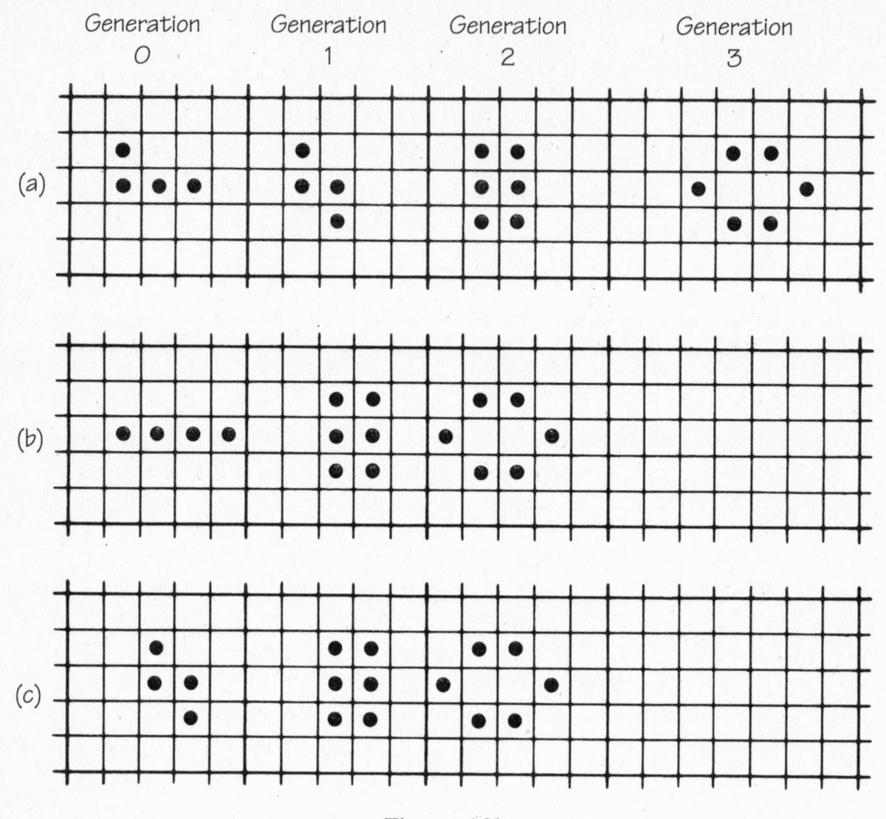

Figure 12b

states (Figure 12c). There were even patterns that one could prove that they do not have any predecessors (these were called "Garden of Eden").

The moral from this game is simple: even with a very simple set of fully deterministic laws, an evolving system can achieve a high level of unpredictability and complexity. In particular, certain outcomes can never be reached, while others have more than one path leading to them. Some patterns achieve a single stable configuration while others send off shiploads of colonists. Small variations in a simple set of initial conditions can lead to very different end results.

As complex as the game *Life* may be, it is still infinitely simpler than real life, of the type we encounter here on Earth. Before I go

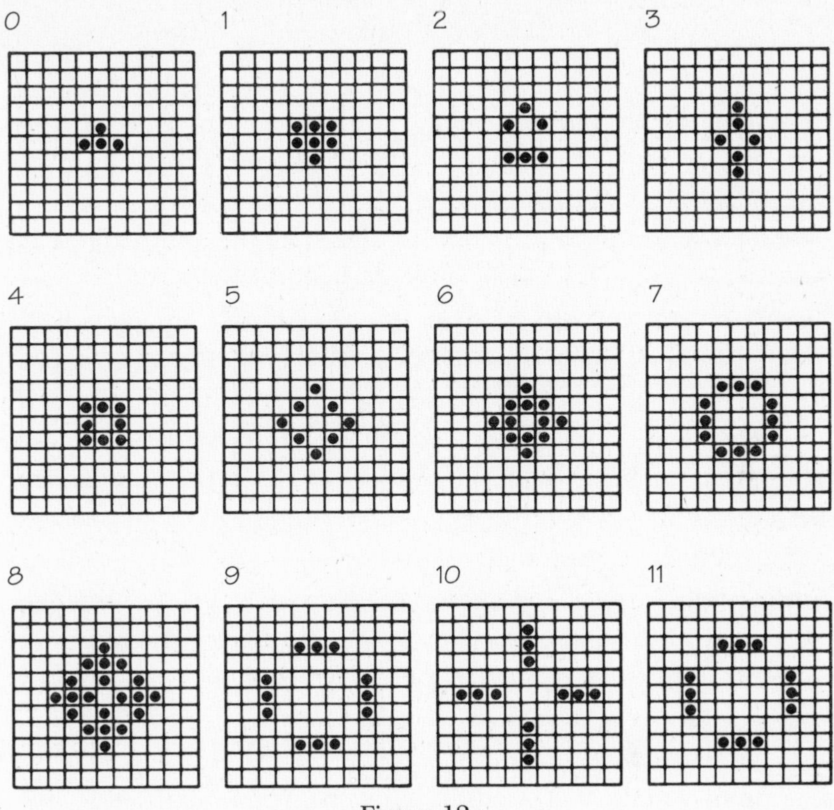

Figure 12c

any further, however, let me first define life. Here I will adopt Carl Sagan's favorite definition: "Life is any system capable of reproduction, mutation and reproduction of its mutations." This definition at least avoids some ambiguities that result from definitions based on functions (i.e., metabolism) or specifically on life on Earth (i.e., the use of DNA). For example, definitions based on ingestion of materials and metabolism could apply equally well to pocket calculators or to airplanes, and definitions based on DNA may exclude life-forms elsewhere.

The living organisms we are familiar with are made of organic molecules, water, the ions of some atoms (atoms stripped of some of their electrons), and a few trace elements. The organic molecules are based on carbon and on its bonds with elements such as hydrogen, oxygen, nitrogen, phosphorus, and sulphur. Water is composed of hydrogen and oxygen. The ions and the trace elements include atoms of calcium, potassium, sodium, iron, chlorine, magnesium, silicon, zinc, cobalt, and a few others (in quantities of less than one-hundredth of 1 percent of all the atoms).

The important point to note is that of all the above elements, only hydrogen emerged from the big bang inferno. All the other elements were formed, via nuclear reactions that involve the strong and weak nuclear forces, in the burning depths of stars or during stellar explosions. Massive stars end their lives in dramatic supernova explosions. In these explosions, stars spill their contents, which include the newly synthesized elements, and disperse them into space, thus altering the composition of the interstellar matter. Similarly, intermediate-mass stars lose their surface layers due to stellar winds and periodic expansions and contractions (pulsations) during the phase in their evolution in which they are enormously expanded. Since the lost envelopes are rich in carbon, such stars serve as a major source of carbon for the interstellar medium. Gravity molds the interstellar clouds into new solar systems, and chemistry (through the electromagnetic force) generates the building blocks for life. We thus see that not only are we literally "stardust," but also that all the four basic forces participate in putting together the stuff life is made of.

But stars are more than just factories of raw materials for life. Stars provide a steady supply of billions of years of energy for their accompanying planets, and this reliable long-lived energy source is apparently required for the development of "intelligent" life. What is the origin of this tremendous energy reservoir, and what is it that determines the longevity and life histories of stars?

The Secret Lives of the Stars

What we normally call *stellar evolution* really represents stellar aging. Stars are born via the gravitational collapse of large gas clouds.

They undergo a variety of identity transformations during their lifetime and eventually they die—namely, become invisibly dim. Thus, in stellar evolution, the equivalent of Darwinian species evolving from one into the next are merely different stages in the life of the star. Because stellar lives are very long compared to human lives, even the entire accumulation of observations of stars throughout human history amounts only to a snapshot in the lives of the stars. Consequently, the deciphering of the processes involved in stellar evolution is similar to an understanding of the process of human aging gained through an examination of a large group of people of different ages.

The basis for stellar evolution is the fact that *the star spends its entire lifetime battling against gravity.* In the absence of a resisting force, gravity would cause a catastrophic collapse of the star to its center. During most of the star's life, the star manages to sustain a neat balancing act in which the pull of its own gravity is precisely counteracted by an outward pressure of its internal hot gas.

The high pressure inside stars results from the high temperature that is generated by *nuclear reactions.* This temperature is approximately 15 million Kelvin (or 27 million degrees Fahrenheit) at the center of our own sun. As protons move frantically in the high-density, high-temperature environment of the stellar center, they manage to collide and feel each other's attractive nuclear force, in spite of their mutual electrical repulsion. Most of the star's energy originates from a nuclear process that fuses four protons (basically hydrogen atoms) into a helium nucleus, which is composed of two protons and two neutrons. In the process, two positrons and two neutrinos are also emitted, but most important, energy is being produced as the protons bind to form the helium nucleus. The mass of the helium nucleus is slightly smaller than that of four protons, and most of the difference is converted to energy according to Einstein's well-known equivalence between mass and energy (energy equals mass times the square of the speed of light). There is hope that in the not too distant future the same reaction will provide an almost inexhaustible, clean source of energy for humanity, when controlled fusion reactions are mastered.

As the star ages, it may exhaust a particular nuclear fuel (i.e., hydrogen) in its core. Such phases are accompanied by contraction of the core due to gravity, and if the temperature becomes high enough, a new fusion reaction involving a heavier nucleus may ensue.

The most important factor in determining a star's evolution is its mass. In general, the *more massive* the stars the *shorter* they live. This is sometimes humorously referred to as "the Nissan and the Cadillac paradox." The Cadillac has a much larger fuel tank, yet the Nissan will get you farther, because the fuel consumption rate of the Cadillac is so much higher. Similarly, massive stars have more fuel (hydrogen) to burn, but they burn it so much faster that they wind up living a much shorter (but more "interesting") life. Thus, while a star like the sun can live for about 10 billion years, a star ten times more massive lives only for about 35 million years.

Objects that are lighter than about 8 percent of our sun's mass (the latter is known as a *solar mass*) never become hot enough to even ignite the nuclear fusion reactions. These are the objects for which astronomer Jill Tarter coined the term *brown dwarfs*. Stars that are more massive than 8 percent of the sun's mass, but less than about 80 percent of a solar mass, have spent their entire lives since birth fusing hydrogen into helium. Such stars can continue to do so for longer than the current age of the universe; hence, these stars have experienced almost no changes since their births.

The next dividing line in mass, resulting in a different evolutionary path, occurs at about eight solar masses. Single stars up to eight solar masses end their lives with a whimper, while those between eight and a few tens of solar masses end them with a bang.

All stars spend most of their lifetime burning hydrogen into helium at their centers. This stage in stellar lives is known as the *main sequence phase*. When hydrogen in the stellar core is exhausted and the energy source is about to be extinguished, the core starts contracting due to gravity. At the same time, the star's outer layers (its envelope) are expanding a hundredfold, and the star is transformed into a bright but cool *red giant.* When this happens to our own sun, about 5 billion years from now, the earth will very narrowly escape being engulfed by the sun. Even so, scorching of the earth and of all life-forms on it is unavoidable. An old joke says that during a lecture in which this end of life on Earth was presented, somebody from the back seats asked in panic: "When is this going to happen?" "Five billion years from now," answered the lecturer. "Ah," came a sigh of relief. "I thought you said five *million* years from now!"

As the giant's core contracts, it becomes hotter and denser, which allows for the ignition of the next nuclear fuel—helium. Two helium nuclei collide and form an extremely short-lived beryllium. During its fleeting existence, this beryllium manages nevertheless to collide with yet another helium nucleus, thus forming carbon, which is necessary for life as we know it.

Another element that is vital for life, oxygen, is formed by fusing one more helium nucleus with a carbon nucleus. Stars less massive than eight solar masses do not go much beyond oxygen in the production of elements. Once the core of the star is composed of carbon and oxygen, the core starts contracting again (remember that gravity is always there!), and the outer layers expand to gigantic dimensions. The star's radiation is so intense that it pushes the envelope off to form a planetary nebula—a diffuse fluorescent cloud of gas. Carbon and oxygen that were mixed into the envelope are now dispersed into the interstellar medium, from which new stars, planets, and life can eventually form. The condensed carbon-oxygen core is left behind as a *white dwarf.* White dwarfs manage to avoid further collapse not by nuclear reactions but rather by a quantum effect. Electrons are prevented from being too densely packed by a quantum mechanical effect known as the *Pauli exclusion principle.* The exclusion principle does not allow more than two electrons to occupy the same energy state. Consequently, electrons are forbidden from crowding the lowest energy possible. Rather, they are forced to occupy higher energy levels, which results in an outward pressure that balances gravity.

Single white dwarfs evolve slowly into obscurity as they radiate the leftovers of their heat. On the other hand, if they happen to have a sufficiently close stellar companion, then mass transfer from the companion to the white dwarf may result. In some cases, this feeding will grow the white dwarf to the critical mass it can support (40 percent heavier than the sun; known as the *Chandrasekhar limit*), and a supernova explosion will ensue. These are the supernovae known as Type Ia, which were used for the determination of the acceleration of the cosmic expansion. New elements, and iron in particular, are synthesized during the explosion. Since the white dwarf is totally disrupted in the explosion with its interior being dispersed into space, all the synthesized elements enrich the interstellar medium. From this medium, later generations of stars and planets can form.

Stars more massive than about eight solar masses live a short but exciting life. These stars still spend most of their time fusing hydrogen into helium in the core. Once the hydrogen is exhausted, they proceed to fuse helium to carbon and oxygen. When they run out of helium, too, the core contracts again, until the temperature becomes high enough to ignite a heavier nuclear fuel yet. Carbon nuclei fuse together to form neon and magnesium, while oxygen nuclei fuse to form silicon. This process repeats itself until the core is composed mainly of iron nuclei. Above the iron core lies an onion shell structure of the "ashes" of all the previous fusion reactions, with hydrogen at the outermost shell and successively heavier elements underneath.

Iron is the most stable nucleus in nature; its protons and neutrons are more tightly bound together than those of any other nucleus. While the fusion of lighter nuclei releases energy, there is no energy to be gained from fusing iron into yet heavier nuclei. In fact, such a fusion costs energy. Consequently, when the iron core of a massive star contracts under gravity's omnipresence, the star cannot be saved by an additional ignition of yet another nuclear fuel. The temperature becomes so high (on the order of a few billion degrees) that the radiation is energetic enough to start dissociating the iron nuclei. This means that the radiation is investing its energy in the wrong place—instead of providing the necessary pressure to balance gravity, it becomes busy breaking nuclei. The consequences are dramatic: the core collapses during a fraction of a second! The collapse can be stopped only when matter becomes so dense that the nuclei of adjacent atoms are so closely packed that they virtually touch each other, leaving no gaps. (In ordinary matter, like a water molecule, typical distances between the nuclei of the hydrogen and the oxygen are a hundred thousand times larger than the size of the nuclei themselves.) At those monstrous densities obtained in the collapse, matter becomes so resistant that any additional compression is almost impossible. Consequently, the matter that falls into the core rebounds, as if it hit a brick wall. The sudden rebound is accompanied by a *shock wave,* like the supersonic boom caused by fighter jets. The heat generated by the shock wave propagates outward through the star's outer layers and, aided by the release of huge numbers of energetic neutrinos, causes a gigantic explosion. The entire stellar envelope, of up to tens of solar masses, is ejected into space with speeds of about ten thousand miles per second. These are known as Type II supernova explosions; a painting called *Supernovae,* by the Hungarian-French op art painter Victor Vasarely gives a wonderfully artistic impression of their observational appearance. The explosion disperses all the hydrogen, helium, carbon, oxygen, sulphur, silicon, and other elements that accumulated in the onion shell structure into the neighboring gas. The enormous heat behind the shock wave allows the synthesis of elements that are not formed in the cores of quiescent stars, such as lead and uranium. This is in fact the main factory for all the elements that are heavier than iron.

The energy output of a supernova is comparable to the energy produced by the sun throughout its entire lifetime. If the exploding star was originally not more massive than about twenty solar masses, then the condensed core it leaves behind is called a *neutron star.* This compact object has a mass that is about 40 percent higher than that of the sun, but a radius of only six miles (ten kilometers). Therefore, neutron stars are so dense that one cubic inch of their material has a mass of about a billion tons.

Neutron stars are prevented from total collapse by the quantum mechanical repulsive force of neutrons. This is precisely the same Pauli exclusion principle as in the case of the white dwarfs, only this time it applies to the neutrons rather than to the electrons. The reason that neutron stars are made primarily of neutrons is that as matter is compressed to extremely high densities, the protons are literally forced to "swallow" the electrons that orbit around them, thus transforming them into neutrons.

There exists, however, a maximum mass that a neutron star can have. Beyond about two and a half solar masses, the inward pull of a compact object's own gravitation becomes so large that nothing, not even the repulsive force of tightly packed neutrons, can prevent the ultimate collapse. The precise value of the maximum mass is less certain than in the case of white dwarfs, because it depends on the relatively less understood physics of matter squeezed to more than nuclear densities; however, it is almost certainly less than three solar masses.

Consequently, in stars more massive than about twenty solar masses, gravity wins the final battle, and the ultimate total collapse of the core is essentially inevitable. Such collapses lead to the formation of the objects for which the physicist John Archibald Wheeler coined the term *black hole.* Extremely massive stars may even skip the explosion altogether and collapse directly into black holes. The cores of massive stars are in this sense less fortunate than Lewis Carroll's Alice. When she realized that her rapid shrinkage was caused by the fan she was holding in her hand, she was able to prevent her collapse into a point by throwing the fan away.

Few astronomical objects have attracted more public fascination than black holes. Science fiction books and Disney movies have helped to transform black holes into household concepts. Even though some of the popular treatments contain gross misconceptions, the fascination is fully justified since black holes are truly intriguing objects, even to astrophysicists.

Recall that general relativity interprets gravity as a curvature in space-time. A mass warps space-time around it just like a heavy ball causes a rubber mat to sag. Near black holes, light rays are so strongly bent that they are swallowed by the hole, being unable to escape. In 1965, Roger Penrose, an outstanding researcher in general relativity from Oxford University, developed mathematical techniques that enabled him to show that when black holes form, general relativity essentially breaks down. In the example with the ball resting on the rubber mat, this would be as if the ball punched a hole through the mat. What Penrose was able to demonstrate is that the structure of space-time is not smooth everywhere. Rather, at certain *singularities* (like the hole in the mat) all hell breaks loose. Singularities are points where space-time is infinitely curved, so that quantities such as density or gravity become infinite (a bit like dividing a number by another number that approaches zero). The existence of singularities in any theory usually signals the breakdown of the theory and the need for some new physics. What Penrose proved was that rapid collapse necessarily leads to a black hole singularity.

In addition to their importance for the theory of general relativity, black holes also provide wonderful examples of reductionism and unification, and we have already encountered such an example with Hawking radiation. Recall that the entropy of a system is a measure of the amount of disorder in the system, and it is related to the number of possible states the system can be in. For example, there are many more ways in which sand could be scattered on the beach than for this sand to be a part of a sand castle. Thus, the entropy of the latter state is lower than that of the former. The second law of thermodynamics says that entropy can never decrease—the amount of disorder in any isolated system stays constant at best, but otherwise it increases. Black holes have a similar property; the area of their horizon (from inside of which no information can escape) never decreases. For example, when a black hole accretes mass or merges with another black hole, the horizon area always increases. The similarity in the behavior of the area of a black hole and entropy has led to the realization that, from a thermodynamic point of view, black holes actually behave like normal systems. The Israeli astrophysicist Jacob Bekenstein used this analogy to reach two remarkable conclusions. The better-known one (which I have already mentioned) is the suggestion that if black holes have an entropy associated with them, then like any other thermodynamic system, they are also characterized by a temperature. Hawking's suggestion that black holes are not absolutely black, but rather emit Hawking radiation, was based on this notion of a black hole temperature. Bekenstein's second discovery is somewhat less widely known but is truly beautiful. First, he showed that the entropy of a black hole cannot exceed a certain maximal value that depends on the size of the hole (the horizon radius) and on its energy. He then went on to show that the same limitation precisely applies to any physical system—namely, that the maximum entropy a system can have is determined by the system's size and energy in precisely the same way that it is determined for a black hole. Finally, recall that the entropy is related to the amount of stored information. Thus, Bekenstein proved that the limit he obtained translates into a maximum rate at which information may be transferred (e.g., by a computer) with no errors. Incredible as this may sound, Bekenstein used the physics that he deduced from his work on black holes to derive a limit on how fast certain computers can operate! If anyone still needs proof that the same laws apply to very different phenomena, I hope that this last example convinces even these antireductionism diehards.

As I mentioned before, some scientists speculate that black holes may even provide gateways to other universes. In this scenario, the collapse that forms one black hole gives birth to a new space-time, and therefore a new universe, that is completely disjointed from our own.

In some respects, black holes resemble the Cheshire Cat that Alice meets during her adventures in Wonderland. The cat disappears from view, leaving only its grin behind. In describing Alice's reaction to this disappearing act, Lewis Carroll writes: "Well! I've often seen a cat without a grin," thought Alice; "but a grin without a cat! It's the most curious thing I ever saw in all my life!" Black holes disappear from view in our universe, but they leave their "grin," their gravitational signature, imprinted in our space-time. Like the Cheshire Cat, these are among the most curious things we have ever seen.

So, what have we learned from our continuously increasing understanding of the lives of the stars? Stars are the grandmothers of life; since the first stars formed, perhaps when the universe was a few hundred million years old, generations of stars that lived and died enriched the interstellar medium with the elements necessary for life. In the relatively thick and cold environments of interstellar clouds, where dust (minuscule grains of carbon and silicate) can provide shielding from ultraviolet radiation, the atoms can combine to form molecules such as water, carbon monoxide, ammonia, and more than a hundred others. Even simple amino acids (like glycine), the building blocks of proteins, have been tentatively discovered near the center of our galaxy.

When I ate my cereal this morning I looked at the side of the box, which gives the "Nutrition Facts." In addition to a variety of carbon-based organic compounds such as proteins and vitamins, the box listed the following as ingredients: iron, calcium, phosphorus, magnesium, zinc, sodium, potassium, and copper. None of these would have been here if it weren't for the nuclear furnaces of stars. Earth, as a part of the solar system, has inherited all of these elements from generations of stellar ancestors. But there is no doubt that we humans "owe" the stars even more. In emphasizing the role that stars played in synthesizing the elements from which the earth, and then life, formed, and also in supplying the energy for life, I do not want to diminish the psychological/poetic role of the stars in human lives. After all, the curiosity to understand the universe and the search for beauty stem precisely from these psychological effects. The last lines in *Dante's Inferno* (in Canto XXXIV) read:

> My guide and I began that hidden route
> to journey back towards the shining world
> not caring now to take a moment's rest.
> We climbed, with him in front and me behind,
> till through a rounded aperture I caught
> some glimpse of those delights that heaven holds.
> And we emerged, once more to see the stars.

A Lapis Lazuli Planet

Even if stars are able to manufacture all the ingredients that are necessary for life (as we know it), this still does not necessarily mean that life will actually emerge. Life also needs a "home"—a suitable planet—and "luck"—that the odds for the appearance of life will be nonnegligible, once conditions are optimal. To identify the main issues that are involved in the emergence of life in general, and of intelligent life in particular, we are forced to examine the one and only example of intelligent life we have encountered so far.

Seen from space, the blue-white appearance of the earth looks very much like a ball made of the semiprecious stone lapis lazuli. As a gas cloud collapses gravitationally to form a star, the centrifugal force leads to the formation of a disk (a protoplanetary disk) of gas and dust around the star. Earthlike planets probably form from the aggregation of dust particles in the star's protoplanetary disk. The Russian mathematician and geophysicist Otto Schmidt was the first to propose, in 1944, that planets formed gradually, by coagulation of dust grains into small planetesimals and the subsequent accretion of such lumps onto other planetary embryos. This scenario gained considerable support during the 1960s, when studies of the moon by the Apollo space program showed that the craters (which were known to have been caused by impacts) were very numerous some 4.5 billion years ago (during the intense accretion phase), but their number quickly declined afterward. The results of Schmidt's work were published in 1969 in a highly influential book, written by one of the leaders of this research, Victor Safronov. The impacts onto the young earth produced enough heat to melt the interior at depths between about one hundred and three hundred miles, thus generating a *magma ocean* under the surface of the earth, which was responsible for intense volcanic activity in the early life of the planet. These volcanic eruptions, which were also accompanied by a bombardment by planetesimals from space, rendered the earth inhospitable for life shortly after formation.

In general, much of the information on the earth's history comes from geology. Geological times are determined by the use of radioactive dating. This method uses the decay of unstable radioactive isotopes to determine the ages of rocks. In radioactive decays, one isotope transforms into another. This process usually happens by a neutron decaying into a proton and an electron, or by a heavy nucleus emitting a helium nucleus. One key property of radioactive decay is a precise *half-life;* namely, starting with a given number of nuclei of a certain isotope, after a time interval of a half-life we will be left with one-half of the original number. Consequently, an accurate determination of the concentrations of certain isotopes can determine the age of the rock containing these isotopes. Among the many available radioactive clocks used in this type of dating, the decay of uranium into lead has proven particularly useful. This was the timekeeping device that enabled Claire Patterson of the California Institute of Technology to establish in 1953 an age of about 4.6 billion years for the earth. Information about the appearance of continents has been gathered, for example, by searching for the mineral zircon. Because of its durability and the fact that it cannot be dissolved by erosion, the presence of zircon testifies to the presence of an earth crust. William Compston of the Australian National University in Canberra and his team discovered zircon in western Australia that places the emergence of continents between 4.1 and 4.3 billion years ago.

From the point of view of the development of life, an important event in the earth's history is the formation of the earth's atmosphere. According to radioactive dating work by Claude Allègre of the University of Paris, the earth began to retain its atmosphere 4.44 to 4.41 billion years ago. Researchers of the earth's atmosphere have little doubt that this gaseous envelope was formed by a process called outgassing—the release of gas from the earth's interior. Measurements of the concentrations of gases such as argon and xenon reveal that most of the outgassing occurred within a million years, some 4.4 billion years ago, but the rest (about 15 percent) continued to be released afterward.

The primitive atmosphere contained very little or no oxygen. It was rich in carbon dioxide, hydrogen, and nitrogen, with smaller amounts of water, methane, sulfur dioxide, ammonia, and a few other molecules. This initial lack of oxygen was a lucky occurrence, since it proved instrumental in the formation of the first organic compounds that were necessary for life to emerge. Already in the 1930s, Alexander Oparin in Russia and J. B. S. Haldane in England realized that the fact that oxygen is able to rob compounds of their hydrogen atoms is destructive to the formation of complex organic molecules. Precisely how the earth's atmosphere evolved is still a matter of considerable debate. In addition to numerous inorganic geochemical processes, the atmosphere's evolution also involved the influences of primitive life-forms. At any rate, an examination of a number of minerals such as iron oxide led Heinrich Holland of Harvard University to conclude that the atmosphere contained very little oxygen until about 2.3 billion years ago. The oxygen concentration started to rise abruptly about then, probably due to photosynthesis by various microorganisms, reaching its present level about a billion years ago. A discovery in 1999 supports this general picture. Roger Summons of the Australian Geological Survey Organization in Canberra and his colleagues discovered fossilized byproducts of the earliest known oxygen-producing organisms (blue-green algae). Their evidence lay nestled in sediments in western Australia dating back about 2.5 billion years.

While interfering with the formation of the initial building blocks for life, the presence of oxygen was, somewhat ironically, absolutely crucial for the appearance of life on land. In the absence of oxygen, ultraviolet radiation from the sun is lethal to life's molecules, such as DNA. Only after a sufficient concentration of oxygen built up in the atmosphere could ozone (which blocks this radiation) form. The ozone molecule is made of three oxygen atoms, and it absorbs ultraviolet radiation very effectively. The explosion of life-forms on Earth, which was manifested by the evolution from single-celled organisms to multicelled ones, probably could not have occurred without the buildup of the protective ozone layer less than a billion years ago.

The history and properties of life on Earth teach us (at least) three extremely important lessons. First and most important, in spite of the wealth of life-forms, there is really *only one life on Earth.* For example, of all the amino acids that exist, life is based on a particular subset of twenty. Furthermore, amino acids come in two forms, which are mirror images of each other in terms of the relative positions of molecular groups. One form is called left-handed and the other right-handed. While both handednesses are found in nature, all the life-forms on Earth use only left-handed amino acids. Finally and most impressively, all life-forms on Earth, from bacteria to humans, share a common carrier of genetic information, the nucleic acids RNA and DNA, and use the same genetic code—the rules for the sequencing of amino acids of the proteins.

What all of this amounts to is that *all the life-forms we see at present probably evolved from a single progenitor.* While there may have been previous, less successful ancestors not shared by all the present organisms, this common, successful progenitor is known as *life's last common ancestor.* The main goal of all the research into the origins of life is therefore much more ambitious than that of the medieval alchemists. The latter merely tried to find a chemical process that would convert base metals into gold. Origin-of-life researchers are trying to find the chemical processes that turned simple molecules into life!

Charles Darwin himself believed in a chemical origin for life (a view he expressed in private correspondence)—namely, that life emerged simply out of chemical processes with no need for divine intervention. However, his "official" statement in *The Origin of Species* still referred to "creation" of the common ancestor.

The commonality of RNA and DNA and of the proteins strongly suggest that the last common ancestor used RNA and DNA as the mechanism for the storage and transfer of genetic information and the proteins for many of the reproduction reactions. Consequently, the life-from-chemistry question can be rephrased as: how did the first nucleic acids and proteins appear?

In attempting to answer this question however, origin-of-life researchers encounter one of those situations that were called "strange loops" by scientist-author Douglas Hofstadter in his fantastic book, *Gödel, Escher, Bach: An Eternal Golden Braid.*

The problem is that present-day experiments are able to synthesize nucleic acids *only in the presence of proteins* and proteins *only with the help of nucleic acids.* As Hofstadter pointed out, this situation is expressed very imaginatively by the artist M. C. Escher's 1948 creation, *Drawing Hands.* In this famous drawing each hand draws the other. Since both nucleic acids and proteins have very complex structures, it appears highly improbable that both could have emerged simultaneously at the same place via chemical processes. However, since the formation of one requires the presence of the other, this strange loop presents a major stumbling block to a chemical origin of life. A way out of this conundrum was proposed independently in the late 1960s by Leslie Orgel at the Salk Institute in San Diego, Carl Woese at the University of Illinois, and Francis Crick (who in 1953, together with James Watson, discovered the three-dimensional structure of DNA) at the Medical Research Council in England. The scenario that Orgel, Woese, and Crick suggested relies on a special role for the RNA, and later it became known as the RNA World. In this model, the prelife RNA was assumed to appear first and to have *both* catalyzed all the reactions leading to replication and synthesized proteins from amino acids. The main reasons RNA was the one charged with this heavy task of being first (rather than DNA or proteins), were that (1) it is easier to synthesize than DNA, and (2) no one could find a way for proteins to replicate without nucleic acids.

Research during the past two decades provided considerable support for the RNA World hypothesis. In particular, Thomas Cech of the University of Colorado at Boulder and Sidney Altman of Yale University discovered catalytic agents (called ribozymes) made of RNA (suggesting that perhaps RNA could indeed catalyze the replication reactions). This discovery is of great significance since all the previously known catalyzers (enzymes) were proteins. Also, a number of studies demonstrated that in special environments, RNA is capable of altering its properties (e.g., develop resistance to breaking by certain enzymes), suggesting that the RNA in the RNA World might have had different properties than present-day RNA.

While many questions remain open, especially regarding the origin of RNA itself, the origin-of-life research is advancing along similar steps to the ones cosmology took in the past few decades. Namely, the questions address progressively more fundamental processes, and ones that took place at earlier and earlier times. It now appears quite likely that RNA did indeed appear first, and that it led to the evolution of DNA and to the synthesis of proteins. All of these elements then combined to become life's last common ancestor.

As I noted above, there are two other lessons from the history of life on Earth, in addition to the fact that there is only *one* life. The second is usually stated in the form of a conclusion: *given the right conditions,* the *initiation* of life is not difficult, and perhaps it is even *inevitable.* There are two pieces of information that have been used as supporting evidence for this claim. First, it appears that life emerged essentially as soon as the earth cooled, settled down, and stopped being bombarded. Remember that the earth only started retaining an atmosphere and creating a core about 4.4 billion years ago, and continents only emerged about 4.2 billion years ago. Bombardment by comets and meteorites (which most likely inhibited the appearance of life) was probably quite intense during the first half a billion years, and the oldest rocks in fact date back to about 3.9 billion years. However, there is indisputable evidence for the existence of life, in the form of cellular fossils of dense colonies of bacteria, that is 3.5 billion years old (at most a few hundred million years after the cessation of the bombardment). Furthermore, Steven Mojzis and Gustaf Arrenius of the University of California at San Diego recently presented evidence for the possible existence of life (in Greenland rocks) as early as 3.86 billion years ago.

The second piece of information that inspired the claim of an easy start for life is the results of origin-of-life experiments. The most famous of these is an experiment by Stanley Miller (then a graduate student) and Harold Urey at the University of Chicago in 1953. They created in a flask an "atmosphere"—a mixture of gases similar to what the prebiotic earth's atmosphere might have been. Their mixture contained hydrogen, methane, ammonia, and water vapor. Electrical discharges intended to simulate lightning were applied to the flask, and the products of the resulting reactions were dissolved in water (simulating the primordial oceans). The net result of this atmosphere in a flask was the production of many amino acids, including some of the types that build proteins. Another experiment, by Juan Oró at the University of Houston, even produced adenine—one of the bases in RNA and DNA. The notion that the Miller-Urey experiment, as well as other similar experiments, simulated successfully the chemical processes on the primordial earth gained some support from the composition of a meteorite that hit near Marchison, Australia. The meteorite was found to contain similar concentrations of the same amino acids that were produced in the Miller-Urey experiment.

Some researchers are not entirely convinced by these arguments, and they point out the following possible loopholes. First, the presence of even small amounts of oxygen in the prelife atmosphere would have seriously hampered the production of the amino acids. There exist serious doubts whether the earth's atmosphere was ever of the composition assumed by Miller and Urey. At present, it is generally thought that the Miller-Urey atmosphere had too little carbon dioxide and too much methane. Second, the conclusion about the easy initiation of life rests on the assumption that all the prelife (prebiotic) chemical processes took place on the earth's surface. In principle, however, it is possible that some of the ingredients of life had in fact been delivered by bombarding comets, meteorites, and dust particles. If this were indeed the case, then it may no longer be possible to claim that the chemistry worked extremely fast. Furthermore, results reported in 1996 by the German biochemist Günter Wächtershäuser (who incidentally works as a patent attorney) suggest that organic molecules that could have been the precursors to life may have first formed around volcanic cracks in the ocean floor. If this is true, then life may not have started as immediately as has previously been presumed, since the chemical processes could have proceeded at a more leisurely pace even in the presence of heavy meteor bombardment.

The third lesson from the history of life on Earth is that while the initiation of life may be easy, it appears that the development of intelligent life is very difficult. Even though life on this planet started more than 3.5 billion years ago, it did not evolve past unicellular animal organisms for about 3.0 billion years. At least two factors contributed to this long delay in the appearance of multicellular animals. One was the lack of atmospheric oxygen (and ozone) and the accompanying lack of protection against ultraviolet radiation. The other was the need to achieve a critical structural level of unicellular life. For example, it was during this delay that cells with nuclei (eukaryotic) developed from cells without nuclei (prokaryotic). The entire wealth of life-forms we see today is the consequence of a main explosion (known as the Cambrian explosion) in emerging life-forms that occurred about 530 million years ago and lasted only about 5 million years (although a few minor pulses started about 600 million years ago). But not only did complexity take a long time to develop, the subsequent evolution is marked by a chain of unpredictable and sometimes pure chance events. For example, among the many fossils found in the Middle Cambrian Burgess Shale fauna of Canada, only one swimming creature possessing a notochord (the longitudinal structural element later replaced by the vertebral column) was found. Had this single member of the chordate phylum not survived, there would have been no vertebrates (and no humans) today.

A second, and better-known, example of chance occurrences that shaped life's pathway are mass extinctions. Paleontologists David Raup, J. J. Sepkosky, and David Jablonski of the University of Chicago showed that since the Cambrian explosion there have been five major mass extinctions and quite a few more minor ones. In the last large extinction, at the Cretaceous-Tertiary boundary 65 million years ago, the dinosaurs perished, and this paved the way for the dominance of the tiny (by comparison) mammals. In 1979, Luis and Walter Alvarez presented strong geological evidence that the Cretaceous-Tertiary extinction was caused by the impact of an astronomical body—a comet or an asteroid—six miles in diameter. The discovery of a crater off the Yucatán peninsula in Mexico, which appears to have a consistent size and geological age, strongly supports the extinction-by-impact hypotheses (of course, many meteorite impacts had been known much earlier). Furthermore, in July 1994 we had the luck of witnessing directly the rare occasion of an impact of a comet onto a planet. Comet Shoemaker-Levy 9 was broken into two dozen fragments while in orbit around Jupiter, and all of these fragments impacted onto Jupiter's atmosphere in mid July. Every impact produced a rising plume similar to a nuclear mushroom cloud, and the "scars" produced at the impact sites could be seen on Jupiter's atmosphere for months after the collisions. I still remember vividly the evening on which the first impact occurred. Gene and Carolyn Shoemaker and David Levy, who first discovered the comet, were at the Space Telescope Science Institute, as were astronomer Hal Weaver, who observed the "string of pearls," as the chain of orbiting fragments was called, and many other planetary scientists. Heidi Hammel from MIT was granted Hubble Space Telescope time to observe the collision that was about to occur, but at that point it was not clear at all whether anything of interest would be seen, since theoretical models showed that unless the fragments were relatively large, the observable effects would be minuscule. The Shoemakers and Levy were holding a press conference in the institute's auditorium, while Heidi, Hal, and other astronomers (myself included) gathered around a computer screen in our operations room to see the first images as they would be transmitted from the telescope. Suddenly, a tiny bright dot appeared above Jupiter's horizon. Heidi asked: "Does anybody know if a moon should appear here?" and a few people started to check for that. However, a short time later it became clear that this was not a moon. A bright plume became clearly visible above the horizon. It was unbelievably beautiful. Somebody took a picture of all of us crowded around the computer screen, with a look of absolute amazement on our faces. It is a fantastic picture of the awe of discovery. The first time I saw that picture I immediately remarked that it is incredibly reminiscent of Rembrandt's painting *The Anatomy Lesson of Dr. Tulp*. In that painting, too, Rembrandt concentrates not on the operation itself, but rather on the astonished, eager-to-learn faces of the anatomy students.

The bottom line concerning impacts is that asteroids and comets surely smashed into Earth in the past, and they were probably responsible for at least some of the mass extinctions, although other causes, such as climatic changes perhaps associated with volcanic activity, might have also played a role. It is therefore clear that humans did not appear as an inevitable outcome of evolutionary theory, but rather as a contingent outcome of many linked events. Had some of these events occurred differently, the alternative pathway might have led to a rather different final product.

But does this necessarily mean that intelligent life is extremely rare in the universe? It is always dangerous to draw conclusions on the basis of one example; therefore, at this point it is virtually impossible to know how general is the applicability of the clues provided by the history of life on Earth. Keeping this serious caveat in mind, the one precious example of life we have is characterized by the following properties:

1. There is really only one "life" on Earth.
2. Life on Earth may be as old as the conditions allowed, and this may indicate an inevitability for the origination of life by chemical processes. However, this conclusion assumes that all the building blocks for life were formed locally on the earth's surface, rather than being brought to Earth by impacts or being formed near deep-ocean volcanic vents.
3. The appearance of complex multicellular life on Earth took a long time. The emergence of humans appears to be at least to some extent a fortuitous outcome of a series of chance events.

We do not know yet whether theories of life satisfy the Copernican principle, since we have only one example of it. But life is certainly an amazing demonstration of simplicity or reductionism. After all, if present theories about the origin and evolution of life are correct, then this miraculously complex phenomenon we call life and even the property we call consciousness are the direct consequences of the simple laws of nature. As Carl Sagan once put it: "These are the things that hydrogen atoms do—given 15 billion years of cosmic evolution."

To increase the sample of life in the universe beyond the one example we know is not easy. Concentrated efforts have already been put into three main directions: searches for life in the solar system, searches for extrasolar planets, and "listening" for extraterrestrial signals. These efforts have already produced some spectacular results. It is certainly beyond the scope of the present book to describe all of these searches in detail; thus I will concentrate on only a few relevant topics. I refer the interested reader to three recent books that discuss many of these topics: *Looking for Earths,* by astrophysicist Alan Boss; *Sharing the Universe,* by astronomer Seth Shostak; and *Other Worlds,* by science writer Michael Lemonick.

E.T. Phone Home

The search for extraterrestrial civilizations has an appeal well beyond the realm of scientific research. In that sweet book I mentioned before, *The Little Prince,* the Little Prince parts from the author in the Sahara to be bitten by a snake (which is supposed to allow his soul to return to his own little planet). Prior to his departure he says to the author: "All men have the stars, but they are not the same things for different people. For some, who are travelers, the stars are guides. For others they are no more than little lights in the sky. For others, who are scholars, they are problems . . . But all of these stars are silent. You—you alone—will have the stars as no one else has them—." "What are you trying to say?" the author asked, to which the Little Prince answers: "In one of the stars I shall be living. In one of them I shall be laughing. And so it will be as if all the stars were laughing, when you look at the sky at night . . . You—only you—will have stars that can laugh!" Imagine, indeed, how we would feel had we knew for sure that up there, near a star, one we may even be able to see at night, there exists another civilization.

Clearly the first searches for extraterrestrial life should start in our own backyard, on other planets in the solar system. Finding life on any other planet would not only be of potentially immeasurable importance for the understanding of the origins of life, but it would also move the Copernican principle yet another giant step forward in a new dimension, by further knocking the earth off its pedestal.

In August 1996, NASA scientists Everett Gibson, David McKay, and their team made an astounding announcement—from a careful examination of a Martian meteorite found in Antarctica in 1984 they concluded that they discovered "evidence for primitive life on Mars." This announcement generated such a commotion that even President Clinton referred to it on the day the news was disclosed publicly. In order to support this claim, the team had to prove two things: (1) that the Antarctic meteorite, known as ALH84001 for being the first meteorite found in the Allan Hills ice fields in 1984, indeed came from Mars; and (2) that the meteorite contained convincing signatures of life.

The first part turned out to be the "easier" one, relatively speaking. Through a series of chemical analyses and radioactive dating techniques geologist David Mittlefehldt of Lockheed has shown that ALH84001 was similar in many respects to eleven previously known Martian meteorites, only that it was much older, about 4.5 billion years old (compared to 1.3 billion years). From the historical record of meltings of the rock and of exposure to cosmic radiation, Mittlefehldt was able to establish the life story of the meteorite. ALH84001 was flung off the Martian surface into space by an asteroid impact some 16 million years ago, and it penetrated the earth's atmosphere about thirteen thousand years ago to reach its icy burial in the glaciers of west Antarctica. This much is noncontroversial, and there is very little doubt about the meteorite's Martian origin. The evidence for life, however, required much more effort. Through the extensive work by Chris Romanek and

Kathie Thomas-Kaperta at Lockheed and Richard Zare at Stanford and their collaborators, the following pieces of circumstantial evidence have been painstakingly collected:

1. The rock was found to contain blobs of carbonate rock, which appear to have formed at temperatures that allow for liquid water to exist.
2. The rock contained tiny structures resembling bacteria in the carbonate globules, although they were at least ten times smaller than known bacteria.
3. The rock contained polycyclic aromatic hydrocarbons (PAHs), which could be the decay and decomposition products of living organisms, although not necessarily.
4. The rock also contained crystals of naturally magnetic iron compounds that are known to form in bacteria on Earth.

While none of the above findings constitutes in itself a particularly compelling case for life, McKay, Gibson, and their team felt that the cumulative weight of the existence of all of these pieces of evidence, all packed within one rock and not being the result of terrestrial contamination, amply proved their case. The scientific community at large was less convinced. This is actually a healthy skepticism, to which any extraordinary claim should be subjected. The point here is very simple; many scientists feel that the accumulation of four rather ambiguous (individually) signatures of life does not reach the threshold that is required for such an important discovery to be convincing. For example, a suspicion has been raised that even if the structures in ALH84001 are bacteria, they might have grown in Antarctica. The suspicion is somewhat strengthened by the fact that bacteria-shaped objects that are similar to some of the structures in ALH84001 have also been found in four lunar meteorites collected in Antarctica. Questions have also been raised about the exact temperature at which the carbonate globules formed. Estimates range from 700 degrees Celsius to 0 degrees Celsius. The growing consensus seems to be that the carbonates formed at relatively low temperatures, probably less than 300 degrees Celsius. But even this temperature is much higher than temperatures at which Earth-type life is known to survive. At present, more than three years after the original announcement, and even with more evidence presented in early 1999, the jury is still out on the question of life on Mars. A definitive answer may have to await a mission to Mars that will be able to penetrate deep underground. In any case, if life is ever found unambiguously on Mars, it will be interesting to establish whether it has any potential relation to life on Earth. By this I do not mean that events such as those in Tim Burton's movie *Mars Attacks,* in which little Martians invaded Earth, may have actually happened. Rather, it will be important to establish whether life on Earth might have started, for example, as a result of some primitive life-forms arriving on board a meteorite from Mars some 4 billion years ago.

Finding signs for primitive life-forms on other planets in the solar system would undoubtedly be an extremely exciting discovery, but everybody will acknowledge that finding life, and especially intelligent life, outside the solar system would attain a much higher level on the excitement scale. There are two ways, in principle, to attempt to reach this high goal. One is to gamble on the possibility of success of a major shortcut and to try to capture a direct "hello" signal from an extraterrestrial civilization. The other is to proceed slowly, step by step, first by finding ways to merely *detect the existence* of extrasolar planets in general, then to attempt to detect *Earth-like* planets, then (probably much later) to develop techniques to actually *image* such a planet and determine its spectrum, and only then, on the basis of the deduced conditions on the planet, to attempt to determine the likelihood for the existence of life.

Both approaches are currently being followed, although most of the effort to "hear" E.T. directly is privately funded, rather than being carried out at universities or at NASA-funded institutions.

The most concentrated listening project, called Phoenix, is conducted by the SETI (Search for Extraterrestrial Intelligence) Institute in Mountain View, California. The name Phoenix was inspired by the project's somewhat unfortunate past history, during which a previously planned project lost its NASA funding due to a congressional decision in 1993, only to arise from the ashes in its new incarnation as a privately funded endeavor. The project uses at present mostly the giant radio telescope in Arecibo, Puerto Rico, which measures a thousand feet across. While the immense aluminum dish keeps a fixed gaze on space, the earth's rotation and the mobility of the receivers ensure that about one-third of the sky can be observed. Phoenix is a "targeted" search; namely, one thousand specific nearby stars that are similar to our sun are examined individually. The hope is that alien intelligent civilizations transmit some narrow-band radio signals that can be identified as such (e.g., the strongest signal the earth transmits right now comes from radar systems). The reason that the radio band (rather than, say, X ray) is considered most promising is that it requires less power to transmit radio signals that are, in principle, detectable. Finding a needle in a haystack is a child's play compared to this search. Detecting a true alien signal would be the equivalent of searching in a thousand haystacks, when not only do you not have any idea what it is you are searching for, but in fact the object you are supposed to find may look like hay.

The relatively small odds for success do not deter Phoenix scientists—a bunch of E.T. enthusiasts. The team is led by radio astronomers Jill Tarter and Seth Shostak and physicist Kent Cullers (who, as incredible as this may sound, has been blind from birth), all of the SETI Institute, whose president is one of the founding fathers of the search for extraterrestrials, Frank Drake. Although Jody Foster's character in the movie *Contact* (which was based on Carl Sagan's book) almost certainly represents a mixture of known scientific figures, there is very little doubt that it was modeled primarily after Jill Tarter (who, you may recall, also coined the name "brown dwarf"). What Project Phoenix does at Arecibo is to "listen" (electronically) to almost 2 billion radio channels for each star, spanning the radio band from 1,200 to 3,000 megahertz. About half a day of "listening" is devoted to every star from the list of about one thousand stars (all within two hundred light-years from the sun). Once a signal is received on any channel, a huge experimental and software operation is set into motion, to test for its E.T. credentials. The problem is of course that a signal could be originating from any number of Earth-related sources, such as airplanes, radar stations, amateur radio stations, satellites, and the like. Thus, Phoenix scientists have generated in advance a huge "library" of local signals. As a first step, any signal is compared to the library using a powerful computer, and the signal is immediately discarded if it matches any known local signal. Right off the bat, this typically weeds out about 80 percent of all the signals (there are hundreds of thousands of signals in any "listening" run).

If the signal is found to hold rock steady at one frequency it is also discarded, since a signal from outer space would suffer a Doppler shift because of the earth's rotation. The properties of the few signals that manage to pass successfully all the initial tests are communicated to a second radio telescope, the 250-foot dish in Jodrell Bank, England. The second telescope has first to confirm that it too is receiving the signal (otherwise it is clearly of local origin). Then, the frequencies of the signals measured at the two telescopes are compared with great accuracy. A real signal from deep space would be Doppler shifted (due to the earth's rotation) slightly differently in the two locations, in Puerto Rico and England. Thus, if the expected difference is not detected, the signal is discarded again.

A few signals raised momentarily the blood pleasure of the scientists involved. A signal from EQ Pegasi, a double-stellar system some twenty-one light-years away, passed all of the above tests. At that point the procedure was to point the telescope away from the star, since if it was interference it was likely to be detected there, too. Only this signal wasn't! The next obvious step was to "listen" (point) to the star again, to confirm that the signal was still there. It was! "Jill and I were now out of our seats," says Seth Shostak in describing this event. However, in the next observation *off* EQ Pegasi, the signal did appear—it was probably a satellite, after all.

At the time of this writing Phoenix has monitored about three hundred of its one thousand targeted stars, and E.T. has not phoned yet. Luckily, the lack of success so far does not discourage the SETI Institute scientists one bit, and it is quite likely that even if Phoenix fails to detect any credible signal, even more ambitious and comprehensive searches will be designed. Again in the words of Saint-Exupéry in *The Little Prince:* "For you who also love the little prince, and for me, nothing in the universe can be the same if somewhere, we do not know where, a sheep that we never saw has—yes or no?—eaten a rose . . ."

Planets Galore

The second method to search for extrasolar life is not to take any shortcuts, but rather to progress step by step, building each investigation on the foundations of the previous findings. The first step in this journey has to be the *detection of planets* around other stars. An important question that needs to be answered before such a detection can be claimed is: what is the difference between a brown dwarf and a planet? After all, both are smaller than a star. The main *physical* difference has to do with the formation process of these two types of objects. Brown dwarfs form like normal stars, from the gravitational collapse of a gas cloud. The only difference between brown dwarfs and stars is that because of their low mass (less than 8 percent of the mass of the sun), they are never able to ignite the furnace of nuclear fusion reactions in their cores, and consequently they remain relatively cool and dim.

Planets, on the other hand, form from the aggregation of solid lumps from the dusty disk that surrounds a forming star. Those chunks of matter form objects that are about half a mile in size, which continue to accrete material to form planetesimals, with sizes similar to that of the moon or even larger. The continuing collisions among these bodies eventually build up planets like the earth. In the gaseous parts of the disk, these protoplanets accrete gas on top of their rocky foundation to form planets like Jupiter. In terms of mass, the dividing line between planets and brown dwarfs is somewhat uncertain, but it is in the neighborhood of ten times Jupiter's mass.

By 1990, even though no extrasolar planets had been discovered, the existence of dusty protoplanetary disks around stars had been established unambiguously, largely through the work of Steven Beckwith, then at Cornell University and presently the director of the Space Telescope Science Institute, and Anneila Sargent from Caltech. Spectacular images of these dark dusty disks around young stars in the Orion nebula and in stellar nurseries in the constellation Taurus were later obtained with the Hubble Space Telescope, and were nicknamed "proplyds" (for protoplanetary disks) by astronomer Robert O'Dell (who was also the first project scientist for HST) of Rice University. Thus, there is no question that the raw materials for forming planets are available for many stars. In one case, in fact, that of the star HD141569 in the constellation Libra, a gap was observed in the dust disk, of the type that might have been swept out by a planet.

So how does one find planets around other stars? To actually spot a planet through a telescope would be much more difficult than to hear the voice of an elementary school boy singing a duet with Pavarotti. The star outshines its planets a billion times. Astronomers therefore use the same indirect methods that were employed to discover the dark matter. Any two stars in a binary systems, like two dancers holding hands and swinging in a circle, revolve around their center of gravity. If the two stars are equal in mass, the center of gravity is precisely in the middle between them.

But if one object is much more massive than the other, as in the case of a star and a planet, then the center of gravity is much closer to the center of the massive body. For example, if we consider the Sun-Jupiter system, the center of gravity is just about at the sun's surface. This means that while Jupiter is revolving in a wide orbit, the sun is merely moving along an orbit as wide as its own diameter. Similarly, if a sumo wrestler was holding hands and revolving with a three-year-old girl, the girl would be virtually "flying" in a wide circle while the wrestler would be basically pirouetting in his place. In principle, therefore, the problem of planet detection is straightforward: the spectrum of the star should reveal the Doppler shift corresponding to the small stellar wobble, and the inferred velocity is directly related to the mass of the invisible planet. In practice, however, this is a rather formidable measurement. Typically, astronomers use Doppler shifts to measure speeds that are higher than two thousand miles per hour, while the tiny wobbles induced by planets may amount to no more than about twenty miles per hour. It was therefore pretty clear from the beginning that a truly heroic effort will be needed to detect planets unambiguously.

The first extrasolar planets were discovered in the least expected place—orbiting a pulsar. Pulsars are rapidly spinning neutron stars, the collapsed cores of supernovae (exploding massive stars). These neutron stars possess enormous magnetic fields as a result of the extreme compression that the cores and their original weak field have experienced. This is the same effect as the increase in the speed of the water flow when you squeeze the opening of a hose. Radio waves are emitted from the two magnetic poles of the pulsar. Because the magnetic axis is not precisely aligned with the rotation axis (this is true for the earth, too), the poles rotate around. The radio waves therefore come in pulses, like the light beams from a lighthouse.

The rotation rates of pulsars (and consequently the pulses) are extremely regular; in fact, some pulsars are the world's best-known clocks, exceeding even atomic clocks in their precision. This precision is so astonishing that when these equally spaced pulses were first discovered in 1967 by Cambridge graduate student Jocelyn Bell and her supervisor, astronomer Anthony Hewish, they first thought the pulses might come from alien civilizations. The extreme accuracy of the pulsar clocks makes it possible to detect (by the shift in the frequency) even the smallest wobbles in their motion. In July 1991, British astronomer Andrew Lyne and his group announced that they had discovered a planet only ten times heavier than the earth orbiting the pulsar PSR 1829 – 10. The discovery caught the entire astronomical community by surprise, since no one expected that, of all places, planets would be found near a star that has undergone one of the most dramatic explosions in the universe. Unfortunately, in the American Astronomical Society meeting in Atlanta, which took place in January 1992, Andrew Lyne admitted that he and his group had made a mistake by not accounting properly for the earth's revolution around the sun, and this introduced the spurious "planet" into their analysis. I was in the audience when Lyne gave his talk, a painful presentation of a serious scientific error. When he finished, there was a burst of applause. We all admired the courage and scientific integrity with which Lyne explained in great detail the mistake they had made. But, as luck would have it, less than a week prior to Lyne's admission of his group's erroneous discovery, astronomers Alexander Wolszczan, then at Arecibo, and Dale Frail of the National Radio Astronomy Observatory in New Mexico announced that they had discovered not one but two planets around the pulsar PSR 1257+12. These planets had masses of 3.4 and 2.8 Earths, and they revolved around the pulsar every sixty-seven and ninety-eight days, respectively. In the aftermath of Lyne's unfortunate story, Wolszczan and Frail's findings were subjected to serious scrutiny and were eventually fully confirmed six months later by an independent group of astronomers led by Don Backer from Berkeley. Astronomers therefore had to accept the fact that planets can exist in the apparently hostile environment of a pulsar. The exact process by which these planets form remains somewhat of a mystery, but at least two models remain viable. In one, the pulsar is actually formed with a close *stellar* companion, but this unlucky star is eroded and evaporated by the intense radiation from the pulsar. Planets can then form from the debris of the destroyed star. In the second model, the system starts with a close pair of white dwarfs. Einstein's general relativity theory predicts that in such a system, the two stars are drawn together by the emission of gravity waves. Eventually, the orbit becomes so tight that the lighter of the two white dwarfs is totally dissipated by the gravity of the heavier companion and it forms a disk around the heavy white dwarf. The planets form from this disk in the same way that planets formed in our own solar system. The massive white dwarf accretes material from the disk and eventually collapses (if there is sufficient mass to drive it beyond the critical limit) to form a neutron star. In a paper I wrote with my colleagues Jim Pringle from Cambridge and Rex Saffer, now at Villanova University, we pointed out that this scenario predicts that planets may also be found around massive white dwarfs that did not quite make it all the way to collapse. The observational test of this prediction, however, is more difficult, because white dwarfs are not equipped with the incredibly precise "clocks" that pulsars possess.

Nobody suspects that the planets around pulsars can harbor any life; the conditions near the unfriendly, intensely radiating pulsar are just too harsh. But the discovery of these planets demonstrated for the first time that orbiting objects not much more massive than the earth are not exclusive to the solar system. Furthermore, with more data, Wolszczan found evidence that PSR 1257+12 may have as many as four planets (one with a mass only slightly larger than that of our moon), but a final confirmation of the reality of the fourth planet will have to await the analysis of future observations. A Jupiter-size planet has also been tentatively identified around another pulsar, PSR 1620+26, by astronomer Don Backer.

As I noted before, finding planets around normal stars sounded like a "mission impossible" until about ten years ago. Existing spectrographs (the instruments that record the stellar spectrum) just did not appear sensitive enough, and in particular the errors introduced by standard spectroscopy work were too large to measure the tiny planet-induced wobble.

These apparently insurmountable difficulties did not scare away astronomers Geoff Marcy and Paul Butler of San Francisco State University. Through a series of ingenious improvements to the observational techniques and procedures, and an incredibly complex software development, this team achieved by 1995, after years of work, what can definitely be considered the best planet-finding system and strategy. Surprisingly, however, the first planet around a normal star was found by another team, that of Swiss astronomer Michel Mayor from the Geneva Observatory and his young colleague Didier Queloz. Mayor has specialized for many years in spectroscopic detection of companion stars, and by improving his instrumentation he also reached the accuracy that was necessary for the detection of Jupiter-like planets. Starting in April 1994, Mayor and Queloz planned to observe 142 stars that were all different from the 120 observed by Marcy and Butler, and all appeared to be single stars in Mayor's previous observations. In September of the same year, they observed a star called 51 Pegasi (in the constellation Pegasus; 51 Peg for short), and four months later they were quite convinced that they detected the wobble corresponding to the star being pulled by a planet. However, to be sure they had to await another six months, for the constellation to appear again in the night sky from behind the sun. The new observing run convinced them beyond a doubt that they had hit the jackpot, and in August 1995 they submitted the paper announcing the first detection of a planet around a sunlike star. The mass of the planet was about half that of Jupiter, but it was incredibly close to the star, orbiting it in a mere 4.2 days. This made the planet seven times closer to the star than the planet Mercury is to our sun. Because of these peculiar properties, many astronomers were at first quite skeptical of Mayor's result. Incidentally, Marcy and Butler could have easily detected this planet had they believed that a planet this large would be found this close to the star. However, it took Marcy and Butler only four nights of observations to convince themselves that Mayor was right after all—and a new era in planet search began. By the end of 1995, Marcy and Butler finally came into their own by discovering two planets, one around 47 Ursae Majoris and one around 70 Virginis. Nine long years of work started to pay off.

At the time of this writing, there are now twenty known planets around sunlike stars, and the list is continuously growing. The masses of these planets range from about half of Jupiter's mass to eleven Jupiter masses. Most of these planets were found by Marcy and Butler, the truly relentless planet hunters.

The most dramatic of these discoveries (so far) came on April 15, 1999, when teams from San Francisco State University (led by Butler and Marcy), the Harvard Smithsonian Center for Astrophysics, and the High Altitude Observatory in Boulder, Colorado, announced the discovery of the first *planetary system* orbiting around a sunlike star. The story of this observation resembles very much the discovery of the planet Neptune, or of dark matter. In 1996, Marcy and Butler detected a Jupiter-size planet orbiting the star Upsilon Andromedae, which is about forty-four light-years from Earth. However, they noticed that the variations in the star's speed exhibited an unusual amount of scatter. After analyzing eleven years of data, they were able to confirm that some of the confusing pattern in the data could be explained by the presence of a second planet. But even then, yet another object still seemed to be pulling gravitationally on the star. Eventually, Butler and Marcy concluded (and the same conclusion was reached by the other teams) that the data are consistent with the existence of three planets. One planet, of about Jupiter's mass, is orbiting the star every 4.6 days. A second one, twice as heavy as Jupiter, takes

242 days to complete an orbit. The third, about four times Jupiter's mass, orbits every about 3.5 years. This discovery demonstrated clearly that our galaxy could be teeming with planetary systems.

There exists one other ongoing planet search effort that uses an entirely different technique—gravitational microlensing. I have already described this technique in my discussion of MACHOs. By monitoring millions of distant stars, a few collaborations—in the United States, Poland, and France—find the rare events in which another, closer star crosses the line of sight, thus causing a temporary "lensing" and amplification of the light of the more distant star. The idea is therefore the following: If the lensing star happens to have a planet, the planet will also cause a lensing event of its own. Recall that the mass of the lens determines the amount of bending of the light and therefore the *duration* of the event (not the amplification of the light intensity), with smaller masses causing shorter events. For the most typical lensing events that have been observed so far, the mean duration was about thirty days. The presence of a Jupiter-like planet would have caused an additional peak in the light, lasting for about a day. The exciting element about this technique is the fact that at present it is the only one sensitive enough to detect Earth-like planets around normal stars. Because of the short duration of these events, however, coverage around the globe is required, to allow for continuous observations. In order to exploit this method, Kailash Sahu of the Space Telescope Science Institute and Penny Sackett of the Kapteyn Institute in the Netherlands established the PLANET (for Probing Lensing Anomalies Network) Collaboration. The collaboration uses four telescopes—two in Australia, one in South Africa, and one in Chile. The idea is very simple. As soon as a lensing event is discovered by one of the groups following the MACHOs, they alert the PLANET Collaboration, which starts following the event around the clock, with the hope of discovering the short-duration peak that is the hallmark of a planet. At the time of this writing, the PLANET group has followed more than two dozen microlensing events. They found quite a few anomalies that are still being analyzed, but they have only one candidate for a planet, which is about ten times the mass of Jupiter.

Where does all of this leave us? Since the mid 1990s, we have evolved from a situation in which no extrasolar planets around ordinary stars were known, to twenty such known planets, including one planetary system. Furthermore, earth-mass planets were found around such unlikely *objects* as pulsars, and giant planets were found in such unlikely *locations* as orbits smaller than that of Mercury. These findings demonstrate very clearly the dangers of drawing conclusions and inferences from the previously single known example of a planetary system, our own. The discovery of extrasolar planets also adds direct confirmation to the strength of the Copernican principle—planets appear in fact to be quite common. We are not living in such a unique place in the universe, after all.

The future of planet searches looks very bright; the NASA administrator, Daniel Goldin, regards the goal of producing a direct image of an Earth-like planet orbiting another star as one of NASA's major challenges and desired legacies. His enthusiasm and visionary outlook are inspiring many scientists. The associate administrator for space science, Edward Weiler, who managed an outstanding Hubble Space Telescope program, was himself also the head of NASA's Origins program, the goals of which are to uncover the secrets of star and planet formation, of galaxy formation, and of the origin of life. Thus, we can be sure that in the upcoming years we will witness many new discoveries.

I took this journey into the scientific "meaning of life" because some astrophysicists now believe that life—yes, our own "intelligent" life—may actually play a role in the making of the universe. This line of thinking suggests that our entire observable universe may be some kind of a "cosmic oasis" (to use the words of cosmologist Martin Rees) in which conditions were propitious for intelligent life. According to this anthropic reasoning, a "typical" universe in the ensemble of all possible universes may not satisfy the conditions that are necessary for the emergence of intelligent life.

The laws of physics in such a universe, for example, may not allow molecules to form or stars to shine. As we shall soon see, if this reasoning is correct, then our conclusions about the existence of a beautiful theory that describes the universe may have to change drastically.

Clearly, scientists are not the only ones interested in the meaning of life (even the British satirical group Monty Python had a movie with this title). In 1991, David Friend and the editors of *Life* magazine published a book titled *The Meaning of Life,* in which they collected photographs and brief reflections from 173 poets, scientists, artists, philosophers, statesmen, and everyday people on the street, on the question of "the meaning of life." Interestingly, a part of the reaction of Timothy Leary, the American counterculture leader and psychologist, addresses our role as curious observers. It reads:

> So why are we here?
> We are here to decipher the digital messages from our sponsors.
> We are here to learn how the universe is designed.
> We are here to understand the gods who programmed us.
> We are here to accurately emulate their grandeur.
> We are here to learn the language in which they speak.

But are we here indeed just to understand "how the universe is designed"? Or is it possible that the design itself is somehow influenced by our existence?

9

A Universe Custom Made for Us?

The apparent resurrection of the cosmological constant is not welcomed by many physicists. Neither is the fact that the deduced contribution of this constant to the density of the universe is smaller than its most natural value by some 120 powers of ten. In the minds of some cosmologists, this raises the possibility that the value of the cosmological constant is determined not by some underlying fundamental theory but rather by constraints imposed on our universe by the mere fact that *we* exist—namely, that only a universe with such a value for the cosmological constant is favorable for the emergence of intelligent life. Is it possible that properties of our universe are determined by our existence? Or put differently, is our universe fine-tuned for life?

Fundamentally, the precise properties of our observable universe are determined by the values of a few fundamental constants of nature, which as far as we can tell are *universal constants* (that is, they have the same values everywhere in our observable universe). These constants include the measures of the *strengths* of all the basic forces—gravitational, electromagnetic, and the two nuclear forces—and the *masses* of all the elementary particles (such as electrons and quarks). For example, in our universe today, the strong nuclear force is about 100 times stronger than the electromagnetic force, and the latter is about 10^{36} times stronger than the gravitational force. Similarly, the mass of the proton is 1,836 times larger than the mass of the electron. It is always more convenient to talk about *ratios* of masses (namely, by how much one particle is heavier than another), rather than the masses themselves, because then we do not have to specify if we measure the masses in grams, ounces, and so on.

The values of these constants, together with the basic principles of quantum mechanics (such as the uncertainty principle and Pauli's exclusion principle, which do not allow, for example, electrons to be too closely packed), determine the sizes of atoms, sizes of planets, and even the sizes of two-legged animals on Earth. It is not an accident that we do not find on Earth two-legged animals that are three hundred feet tall. The bones of such an animal would break under the animal's weight, because the strength of the chemical bonds in the bones, which is determined by the strength of the electromagnetic force, would be too weak to hold against gravity's pull. Had the strength of gravity been even weaker than it is, or rather had the ratio of the gravitational to electromagnetic forces been smaller, the average size of animals on Earth might have been larger than it is (not a particularly pleasant thought if we happened to stay the same size). Similarly, the ratio between the strength of the strong nuclear force and the electromagnetic force determines how many different atoms can exist stably in nature. The reason that there are no more than 105 elements in the periodic table is that if you attempt to form a nucleus with even more protons, their mutual electrical repulsion exceeds the mutual attraction provided by the strong nuclear force. Had the ratio between the strengths of the nuclear and electromagnetic forces been somewhat smaller, say, with the strong nuclear force being 50 percent weaker, then even the carbon atom could not have existed, and there would have been no organic chemistry and no us to examine this question.

Similarly, the high value of the ratio of the proton mass to the electron mass (the proton is 1,836 times heavier) allows molecules to be stable. The quantum mechanical uncertainty in the position of a particle is larger the *lighter* the particle. Thus the positions of the relatively light electrons in molecules are much less well specified than those of the nuclei (which are made of protons and neutrons). The nuclei stay put (more or less) and give molecules their well-defined structures and thereby their properties. Had the proton-to-electron mass ratio been much smaller, then molecules would have looked like a blurred image produced by a beginning amateur photographer. In such a case, the very complex molecules of organic matter would not have existed.

The above and other similar "coincidences" between the relative values of the fundamental constants of nature have inspired in some scientists the notion that our universe is particularly conducive to the existence of life. The most widely cited of these coincidences is related to the production of carbon, which is the most fundamental building block of life as we know it. Carbon is produced by a chain of fusion reactions in nuclear furnaces inside stars. However, the final reactions in this chain, which were suggested by the great Cornell astrophysicist Edwin Salpeter in 1952, appear on the face of it to be rather improbable. Salpeter proposed that two helium nuclei combine first to form an unstable isotope of beryllium, and the latter, in spite of its extremely short lifetime (about 10^{-17} seconds), manages to absorb yet another helium nucleus before it disintegrates, to produce carbon. In 1954, Cambridge astrophysicist Fred Hoyle realized that for such an improbable reaction to produce the observed cosmic abundance of carbon, a certain special condition must be satisfied. Hoyle, who was one of the founders of the theory of the synthesis of the elements, predicted the existence of a particular energy state of carbon. He estimated that this state has an unusually high probability of being produced from beryllium and helium (this is known as a "resonant" state). For this formation to happen, the energy of the resonant state of the carbon nucleus has to fit very closely that of the combined energy of the beryllium and helium nuclei. This is a bit like attempting to reach a very specific height with a relatively primitive two-stage rocket. If the combined normal boosts of the two stages add up to give more or less the required height, then the chances of reaching the correct height are better than if they typically add up to a much higher or much lower height. Remarkably, the carbon energy level predicted by Hoyle was subsequently found experimentally. If this energy level had not existed, carbon (and indeed all the heavier elements) would not have been produced, and carbon-based life would not have emerged. Since all the energy levels of a nucleus are ultimately determined (although in a complicated way) by the values of the universal constants, such as the strengths of the nuclear and electromagnetic forces and the masses of particles, the existence of the resonant state of carbon has been regarded by some as "evidence" that our universe is *fine-tuned for life.*

Personally, while I have always found Hoyle's intuition on the resonant level of carbon enviable, I never regarded the existence of the coincidence itself as evidence for a universe that is favorable for life, partly for the following reasons. In 1989, in a work I did with astrophysicists David Hollowell, Achim Weiss, and James Truran, we calculated how much carbon would be produced in stars if we were to shift (artificially) the energy of the resonant state of carbon to a different value. We found that we could change the energy appreciably and still produce substantial amounts of carbon. Furthermore, it is quite possible that if we were to tinker with the constants of nature, so as to change the energy of the resonant carbon state, other states would come into play, with the net outcome of carbon being produced via other paths.

There is another, more fundamental reason why I find the claim that the laws of nature are fine-tuned for life unconvincing. The basic ingredients that are indispensable for life *do not play any particularly fundamental role in the underlying theory that unifies the basic forces and the elementary particles.* For example, in the standard model, we find that electrons, muons, and tauons play an *equal* role; no one is more fundamental than the other. Yet electrons are vital in the atoms for life, while the importance of muons and tauons for us (as an intelligent life-form) is marginal at best (recall: "Who ordered that?"). In a universe that is "fine-tuned for life," one would expect to see a stronger correlation between the quantities that are *significant* for life and their *significance* in the ultimate theory of the universe. The poet Stephen Crane expressed thoughts similar to mine nicely:

> A man said to the universe:
> "Sir, I exist!"
> "However," replied the universe,
> "The fact has not created in me
> A sense of obligation."

Needless to say, the concept of our universe being "fine-tuned for life" also violates the generalized Copernican principle and would therefore constitute a major departure from a beautiful theory of the universe. Physicists would need a very serious reason to embark on this route and abandon the idea of beauty.

Having said that, even if the properties of the universe are not *determined* by the existence of life within it, clearly our universe and the laws and constants that define it have at least to be *consistent* with our existence. Namely, we will surely not discover that the laws of nature and the values of the universal constants in our universe do not allow intelligent life to appear. The question remains, however, whether the values of all the constants are determined by an underlying fundamental theory, or whether some of them could be more or less accidental. If the latter is true, then one could attempt to relate the values of the constants to the existence of intelligent observers. This line of reasoning has become known as the anthropic principle. The logic that underlies this "principle" is somewhat unusual, as the next "scene" will demonstrate.

Newton

A garden with apple trees in a late afternoon hour. Under one of the trees sits a long-haired young man, in his early twenties. He appears to be deep in thought. He holds a quill in his hand and a paper is resting on his lap.

YOUNG MAN [*to himself*]: It feels good to be home. This plague is monstrous, I do hope it will be over soon. At least this rest gives me a chance to think more about light.

[*There is silence again; the young man appears detached from everything in his surroundings, as he is trying to calculate something in his head. Suddenly an apple falls from the tree, right next to the young man. He is startled, as if suddenly awakened from a dream.*]

YOUNG MAN: Good Lord! It is a good thing it did not fall straight on my head.

[*He picks up the apple and looks at it as if he is seeing an apple for the first time.*]

YOUNG MAN: [*to himself, loudly*]: Why is it that apples always fall straight down to the ground? Why don't they fly straight up, or sideways?

[*He sits silently again, deep in thought.*]

YOUNG MAN: Maybe there are places where they actually do not fall down, but rather fly up. Could it be that only on Earth apples are falling down?

[*He is silent again for a few minutes.*]

YOUNG MAN [*in clear excitement*]: Wait a minute! Suppose that there were planets on which apples fly up. In all of those places the seeds of the apples would not get into the ground and apple trees would not propagate! So even if you had a few apple trees to begin with there would not be *any* apple trees in those places after a while. This means that on any planet on which apple trees exist, apples *must* fall down to the ground!

[*The young man takes his quill and after a few minutes of thought writes at the top of the page: "Philosophiae Naturalis Principia Mathematica" (The Mathematical Principles of Natural Philosophy).*]

Clearly the logic in the above fictitious scene does not represent *at all* Newton's real train of thought, nor his conclusions. However, this type of reasoning does represent (admittedly with quite a bit of exaggerated sarcasm) *some* of the arguments made in favor of the anthropic principle.

Anthropics

If you scatter many mouse traps of a certain size in a mouse- and rat-infested cellar, you are not surprised afterward to find that all the mice that got caught are smaller than the size of the trap. This is an example of a *selection effect,* the fact that results of experiments or observations are affected by the methods being used. Similarly, if you observe the skies with a relatively small telescope, you can only expect to see the brightest objects. Before any conclusions can be deduced from astronomical observations, considerable thought needs to be given to identifying all the possible selection effects that might have affected the results of the observations. This is particularly true if *general* inferences are to be made. When carbon-based astronomers observe the universe at large, it would be surprising if they found that this universe cannot produce any carbon. Therefore, it is certainly the case that the universe's observed properties are at some level biased by the selection effect introduced by our own existence. For example, it is clear that we could not observe a universe that was only a million years old (and had a size of just a million light-years), because carbon, which is produced in the depths of stars that live much longer, would not have formed yet in such a young universe. Similarly, we would not have been around to observe a universe in which the density of matter was so low that galaxies never formed.

Modern arguments discussing coincidences in values of universal constants and their possible relations to observational selection effects were presented in the 1950s by physicists Gerald Whitrow and Robert Dicke, but were most clearly formulated in the 1970s by Cambridge relativist Brandon Carter. Carter defined what he called a "Weak Anthropic Principle" and a "Strong Anthropic Principle." The definition of the weak principle reads (I adopt here the definition from the excellent book *The Anthropic Cosmological Principle,* by John Barrow and Frank Tipler):

> The observed values of all physical and cosmological quantities are not equally probable, but they take on values restricted by the requirement that there exist sites where carbon-based life can evolve and by the requirement that the universe be old enough for it to have already done so.

Put simply, the weak anthropic principle is a requirement for consistency: Our pocket universe has to have properties that allow for our existence, as carbon-based life. In the broader context of eternal inflation, and the possibility of an ensemble of universes, the weak principle acquires a more significant meaning, which I will discuss shortly.

The strong anthropic principle states (again I adopt the definition of Barrow and Tipler): "The universe must have those properties which allow life to develop within it at some state in its history." Namely, this principle claims that the laws of nature in *any* universe *must* allow for life to emerge. The strong principle really crosses the borders of physics into teleology. Since there is no hope that a claim of this type will ever be testable by scientific methods, because clearly not all the possible universes are accessible to us, I will not discuss the strong principle any further. I will only note that the strong anthropic principle is in some sense almost orthogonal to the Copernican principle, since it puts life center stage.

As a requirement for consistency alone, the weak anthropic principle is noncontroversial, but it is also not particularly useful. In fact, if applied indiscriminently, it may even be harmfully counterproductive, as the following simple examples will show. Imagine that atmospheric scientists were trying to determine why the earth's atmosphere is rich in oxygen. On the basis of anthropic reasoning they might have argued like this: *We* live on Earth; *we* need oxygen for breathing; therefore, Earth's atmosphere must contain oxygen. Or alternatively, the argument could go: *We* live on Earth; to develop complex life, shielding from lethal ultraviolet radiation is required; ozone, which is a molecule composed of three oxygen atoms, provides such shielding; therefore, the atmospheres of planets on which complex life has evolved must be rich in oxygen. While there is nothing wrong with any of these statements, accepting them as an *explanation* would be a serious mistake, since it would prevent scientists from seeking a true explanation, in terms of physical and chemical processes.

A second example is related to the value of omega, the measure of the cosmic density of matter. In the preinflationary-model era of cosmology, scientists might have used (and in fact a few have used) the following anthropic reasoning: If omega was much larger than one, the universe would have collapsed to a big crunch too soon for stars to form and evolve. If omega was much smaller than one, the universe would have expanded so fast that galaxies and stars would not have had time to form. Therefore, omega needs to be close to one, because these are the only conditions under which life could form. Again, while there is absolutely nothing wrong with any of the above statements, adopting them as the *explanation* as to why omega is equal to one would have been a sad error. The inflationary model (if correct), on the other hand, truly *explains* why our universe should be flat—or in other words, why omega is equal to one—and it does so in the context of a beautiful theory that obeys the Copernican principle and makes no reference whatsoever to *us*.

The question is, then: is there any sense in which anthropic reasoning may mean more than a mere consistency check? Namely, does it have any explanatory power? The answer is that in principle it could have. Recall that *eternal inflation* predicts the existence of an infinite ensemble of universes, in which on every scale you find a pocket universe surrounded by regions of false vacuum that expand dramatically. The laws of physics and/or the values of the universal constants may not be exactly the same in all of these pocket universes. Consequently, in some of these pocket universes life may never form. Some may collapse, for example, before they have a chance to form any stars, or even before they emerge from the primordial, thermal-equilibrium fireball. The latter will not manage to form even atoms. In this sense, one can talk about the subset of anthropically selected universes—those in which the laws and the constants allowed life to emerge. Now, there are two possibilities: either the values of all the constants in any pocket universe are determined unambiguously by some fundamental theory, or some are accidental. In the former case, anthropic reasoning has no place whatsoever. In the latter case, however, the values of some constants that may otherwise appear baffling could suddenly make more sense once one realized that those values are found in an anthropically selected universe. For example, suppose we thought that the value of omega is equal to one in our universe but that this value is not determined by any particular theory. Realizing, however, that an omega either much larger than one or much smaller than one would have resulted in no life would have somewhat quenched our surprise. This example demonstrates also the dangers of anthropic reasoning. Accepting anthropically based explanations too soon, before all the other avenues have been explored, may result in giving up unnecessarily on an explanation based on a fundamental theory.

One of the most serious weaknesses of the anthropic principle is its lack of *predictive* power. Even its strongest proponents usually use anthropic reasoning only post factum, to explain the value of some constant that has already been determined observationally, rather than to predict the results of some future observation. Surprisingly, however, the weak anthropic principle has been used to make one rather shocking prediction, that extraterrestrial intelligent beings are so exceedingly rare that we can essentially regard ourselves as being alone in the Galaxy!

Home Alone?

The argument about the rarity of extraterrestrial intelligent civilizations is due to astrophysicist Brandon Carter, who was also the first to formulate precisely the anthropic principle. This ingenious argument is based on one observation and one assumption. The *observation* is the following: it took a long time, about 4.5 billion years, for *Homo sapiens* to appear on Earth. This time is comparable, to within a factor of two, to the maximum time a biosphere can exist on Earth. The latter cannot be longer than 10 billion years, since this is basically the lifetime of our sun (at least in its present state, in which it fuses hydrogen into helium, and provides for a constant source of energy). After that, as the sun becomes a red giant, it will boil the oceans away and scorch all life on Earth. Carter then made the following crucial *assumption:* he assumed that the average time needed to evolve an intelligent species on a planet and the lifetime of the central star are a priori two *entirely independent* quantities. Put differently, this assumes that intelligent life can emerge on planets at any time with respect to the life of the central star. For example, this assumes that, *in principle,* intelligent life could appear on Earth after a billion years instead of 4.5 billion years. On the face of it, this looks like a reasonable assumption, since the lifetime of stars is determined by nuclear reactions, while the biological evolution time is determined by biochemical reactions and evolution of species. The point now is the following: if two numbers (which can be as large as tens of billions) are truly entirely independent from each other, then chances are that one is much larger than the other. For example, stop for a moment and choose a number between one and 100 billion. What did you choose? I chose 630. There is a very small probability that our numbers are nearly equal. Consequently, Carter argued, in principle, the *average* time it takes to develop intelligent life is either *very short* compared to the lifetime of the star or *very long.* Let us now examine each of these possibilities separately. Imagine that the *average* time for evolving intelligent life is *very short* compared to the stellar lifetime—say, 100 times shorter. Since the sun's lifetime is about 10 billion years, this means that, on the average, intelligent life should appear on a planet around a sunlike star in about 100 million years. In that case, it is very difficult to understand why in the *first* place we found life (the earth-sun system), the two times were found to be *nearly equal.* It was much more probable that we would have found that *Homo sapiens* emerged after only 100 million years, instead of 4.5 billion years. Therefore, the fact that intelligent life took nearly as long as the sun's lifetime to appear on Earth makes it highly unlikely that the *average* time for biological evolution is *much shorter* than the stellar lifetime. Similarly, if the average annual income in a country is $20,000, it is highly unlikely that the first person you meet actually makes $2 million. Let us therefore examine the other possibility, namely, that the average time to develop intelligence is *much longer* than the stellar lifetime, say 100 times longer. Can this explain the fact that it took so long to develop intelligent life on Earth? Note that in this case *intelligent life will simply not develop in most cases,* since biology will die once the star becomes a red giant, and therefore intelligent life will never get a chance to evolve. Given the undeniable fact that *we are here,* which is the essence of the weak anthropic principle, what can we expect to find in the first place that harbors life—the earth-sun system? Clearly we will not find that the time to develop intelligence was *longer* than the stellar lifetime, since intelligence would not have developed after the star's death. We will necessarily have to find the two times to be nearly equal, since this is the longest the biological time can be. But this is precisely in agreement with the observation. The bottom line is that if Carter's argument is correct, the *average* time required to develop intelligent life is *longer* than the lifetime of the star, and therefore life will generally not develop. We, in this case, are the rare exception!

Ever since Carter published this argument, in 1983, I have been extremely bothered by it, not just because I find the idea of being essentially alone in the cosmos depressing. While the argument does not specifically *require* us to be special (and is therefore not in direct conflict with the Copernican principle), its consequences (if true) would make us incredibly special. I am sure that many would find this idea appealing, but from the point of view of the beauty of the physical theory it "smells" anti-Copernican, even though it is formally not. As I will point out later, there are still many ways to make us *feel* special that do not require even a hint of violating the beauty of the fundamental theory.

In spite of my emotional dissatisfaction with Carter's conclusion, I had to admit that it was based on a beautifully constructed argument, and for a long time I was unable to identify any serious loopholes in it. In 1998, I finally discovered and published a potential flaw in the argument. It is important to emphasize that I cannot *prove* that Carter's argument is wrong, but I showed that it *could* be wrong.

The potential mistake is in the assumption that the timescale for biological evolution and the lifetime of the star are *independent* quantities. Recall that the development of complex life on Earth had to await for the buildup of oxygen in the atmosphere and the subsequent establishment of a protective ozone layer. Before that, ultraviolet radiation was lethal to both nucleic acids and proteins. Computer simulations of the evolution of Earth's atmosphere indicate that there were two main phases in the oxygen buildup. In the first, oxygen was released when radiation from the sun broke molecules of water vapor. In general, the duration of this phase is determined by the *intensity of the ultraviolet radiation from the central star,* since this radiation is the one that dissociates the water. The quality of the radiation that stars emit (whether it is mostly visible light, ultraviolet, infrared, etc.) depends on their surface temperatures. Very hot stars emit most of their radiation in the ultraviolet while cool stars emit mostly in the infrared. Our sun, for example, emits mostly visible light, and this is why natural selection resulted in us seeing in the visible, rather than, say, in the ultraviolet. Therefore, the time to release oxygen (which is closely related to the time to develop life) depends on the surface temperature of the central star. However, the surface temperature of a star is related in a very precise way to the star's lifetime. Hot stars are massive and have short lives, while cool stars have low masses and live for very long times. Thus, we find that contrary to Carter's assumption that the biological and stellar times are entirely independent, these two times are actually *related.* In the second phase of oxygen buildup in

Earth's atmosphere, which lasted about 2 billion years, oxygen was mostly produced through photosynthesis by microorganisms. The rate of photosynthesis, however, is dependent on the power received from the star, and is therefore also related to the stellar lifetime. Therefore, the crucial assumption about the independence of the biological time and stellar lifetime may be unjustified. Furthermore, by using a simple (but somewhat technical) model for the dependence of the oxygen level on the stellar properties, I was able to show explicitly that because the biological time increases quite rapidly when the stellar lifetime increases, *the most probable* relation between the two timescales is them being *nearly equal.* Thus, my simple calculation showed that the fact that it took about half the lifetime of our sun to develop intelligent life on Earth may tell us *nothing whatsoever about the frequency of extraterrestrial civilizations.* In fact, in most cases in which an intelligent civilization develops, it may take a considerable fraction of the lifetime of its central star.

My work definitely does not imply that extraterrestrial civilizations exist. Rather, it shows that an earlier conclusion that they *don't* is at best premature. A true determination of the existence or nonexistence of extraterrestrial intelligent civilizations, or extraterrestrial life in general, will come not from speculative statistical arguments, no matter how fancy they are, but from biological and astronomical research of the type I discussed in chapter 8.

We are now finally in a position to attempt to assess whether and how anthropic considerations may change our thinking about the cosmological constant, and the constant's implications for cosmology in general and for the beauty of the final theory of the universe in particular.

Beauty in Distress

The tentative discovery of universal acceleration and the implied contribution of the cosmological constant (or the energy of the vacuum) to omega (total), of about 0.6 to 0.7, raised two important questions: (1) what is it that makes the cosmological constant so small (possibly smaller than its "natural" value by some 120 powers of ten), yet different from zero? and (2) why does the constant start to dominate now? Namely, why out of all the possible times in cosmic history are we living exactly when the contribution of the ever-declining density of matter has been overtaken by the constant energy stored in empty space? In a moment of black humor, cosmologist Mike Turner of the University of Chicago called the latter "the Nancy Kerrigan question," because during the sad incident in 1994 in which ice skater Kerrigan was assaulted, she reacted with a shocked: "Why me? Why now?"

Most attempts to determine a theoretical value for the cosmological constant (lambda) have failed by many orders of magnitude (123 in one model). In despair, some cosmologists have invoked their last resort—the anthropic principle. Basically, the idea is to examine whether the value of the constant appears reasonable, when seen as a prerequisite for human existence. In other words, is it possible that only in a universe with such a constant could intelligent life emerge? Particle physicist and cosmologist Steven Weinberg, who generally objects to anthropic reasoning and believes that all the universal constants should come out of the fundamental theory because they "are fixed by symmetry principles of one sort or another," still admitted in recent years that "the one constant of nature that may have to be explained by some sort of anthropic principle is the one known as the cosmological constant." In a series of works, Weinberg pointed out that if lambda is too large, then the universe embarks on an accelerated expansion before structures like galaxies and eventually humans have time to appear. In fact, as I noted before, once the cosmological constant starts to dominate, any systems held together by gravity that have not formed already will never get a chance to do so. In this sense, it should come as no surprise that in our own pocket universe the contribution of lambda is small. Weinberg and colleagues calculated probable values for the contribution of the cosmological constant to omega (total), given our existence, and found values that are generally consistent with the recent observations. Cosmologists Martin Rees of Cambridge and Max Tegmark of Princeton demonstrated in 1998 that the situation is somewhat more complicated, since the formation of structure depends not only on the expansion rate, but also on the amount of clumpiness that exists when the universe emerges out of the inflationary era. Recall that the COBE satellite found ripples or fluctuations in the density with a magnitude of one part in one hundred thousand, but this particular value still does not have a convincing theoretical explanation in the context of inflation. It is clear that if these initial fluctuations are larger—namely, the clumps are denser—then structures like galaxies can grow faster. Tegmark and Rees showed that if the density fluctuations are too small, less than one part in a million, the structures that form are too dilute and thus may be unable to cool efficiently and form stars. On the other hand, if the fluctuations are too large, more than about one part in ten thousand, the galaxies are so dense that planetary orbits may be disrupted by stellar encounters. Planets would therefore not remain attached to their central stars, and the lack of an energy supply would not allow life to develop. Rees and Tegmark thus concluded that anthropic considerations limit the range of possible values for the density fluctuations and that the value of one part in one hundred thousand is anthropically favored. In a somewhat related work, Jaume Garriga from Barcelona, Alex Vilenkin from Tufts, and I showed in 1999 that if we assume that humans are typical, mediocre observers (i.e., they can be expected to observe the most probable values of the cosmological constant and of the density fluctuations), then the inferred value of the cosmological constant may not be too surprising. We also showed that the time it takes the cosmological constant to dominate the cosmic energy density can be expected to be of the same order as the time it takes intelligent observers to appear on the cosmic scene, thus potentially answering the "Why now?" question.

Many cosmologists are not convinced by these arguments, and

they consider any attempt to invoke the anthropic principle (in any form) as throwing in the towel much too early. The point is that resorting to the anthropic principle implies at some level an admittance that the values of these quantities are not determined by the fundamental theory of the universe, but are rather what they are simply as a result of some accidents of cosmic history. According to this point of view, we happen to find ourselves in one of the universes in which the value of the cosmological constant was favorable for life. As I described before, a similar application of anthropic reasoning to the value of omega (total), while making no erroneous statements, proved entirely unnecessary and *irrelevant* once a theory that *determines* omega (namely, inflation) had been identified.

The value of the contribution of the cosmological constant to the total energy density does pose, however, a frightening challenge to the beauty of the final theory. Even if it is fully determined by the theory, a value like 0.6 to 0.7 looks sufficiently bizarre and unexpected that it is difficult to see at this point how such a value could be a *simple* result of the *underlying symmetry principles*. This is particularly true if the constant has really been reduced from its most natural value by more than 120 orders of magnitude, only to narrowly escape going all the way to zero. Thus, the beauty of the eventual fundamental theory is seriously questioned.

If, on the other hand, the value of the cosmological constant's contribution to omega and the answer to the "Why now?" question are determined *entirely* by anthropic considerations, then this could be a violation of the generalized Copernican principle. The implication would be that one of the most significant properties of our universe, the *dominant* contribution to the energy density, is determined not by an *inevitable* fundamental theory but rather by our own existence. The beauty of the fundamental theory is therefore again facing a huge obstacle.

Does this mean that the final theory of the universe is ugly (according to my definition) after all?

10

A Cosmological Aesthetic Principle?

Beauty in physical theories of the universe is not an afterthought. The German mathematical physicist Hermann Weyl, who contributed greatly to the theory of gravitation, is quoted as having said: "My work always tried to unite the true with the beautiful; but when I had to choose one or the other, I usually chose the beautiful." In fact, there exist at least two known examples, one in relation to gravity and the other to the neutrino, in which Weyl's aesthetic sensibility proved right, in spite of early apparent conflicts with the prevailing wisdom. Thus, the notion of beautiful theories should not be abandoned, at least not without a serious fight.

Are there ways out of the conundrum produced by the recent cosmological observations (if confirmed)? Is there still a beautiful theory lurking in the background and waiting to be revealed, or has beauty reached its ultimate limitations, in the sense that some properties of our universe lie beyond the control of a beautiful theory? Since I have included the generalized Copernican principle in my definition of beauty, proponents of the anthropic principle may argue for the latter possibility. It is certainly the case that Copernican modesty can be applied only to a limit. For example, we would not expect to find ourselves on a planet orbiting a twenty solar mass star, since such stars live only for about 6 million years and therefore complex life would not have had time to develop. Similarly, we would not expect to find ourselves in any universe that would not have allowed complexity to develop. It is therefore not impossible that the particular value that was tentatively determined for the cosmological constant and the fact that the constant started to dominate the cosmic energy density only now are both consequences of the fact that we live in an anthropically selected universe, at an anthropically selected time. This would mean that only universes with cosmological constants of about that value (0.6 to 0.7 of the total contribution to omega) can harbor life and that life exists only during a certain time interval in the lives of those universes, which happens to coincide with the dominance of the vacuum energy density. All of this is indeed possible, but I must say that I personally find this possibility neither very convincing nor particularly appealing. For example, while it is true that life cannot exist in the universe at *any* random time, because of such considerations as carbon production (which does not allow life to appear too early) and dwindling energy resources (through stellar deaths) and proton decay (which do not allow it to exist too late), the time interval that allows the existence of life is nevertheless astronomically long. I therefore still feel that theories that require us to live in a very special time are ugly. Admittedly, the weak anthropic principle has become much more powerful in the context of the eternal inflation scenario than it was in relation to only one existing universe. If there were only one universe, then the weak anthropic principle would have amounted to the rather trivial statement that the properties of this one universe have to be consistent with our existence—not a particularly revealing concept. When one discusses a potentially infinite ensemble of universes, however, with possibly different sets of physical laws and/or different universal constants, then it is clear that only a certain subset of this ensemble may allow complexity and life to develop.

Consequently, cosmologist Alex Vilenkin was able, for example, to calculate probabilities for the realization of certain sets of constants, assuming that our civilization is a "mediocre" inhabitant of the ensemble of universes. In other words, the values of the constants of nature in *our* universe are just about the most probable ones that allow life to exist. Anthropic reasoning may in this case provide some insights into otherwise perplexing aspects of our universe, such as the value of the cosmological constant.

This argumentation may definitely turn out to be correct, but most physicists would like to have to resort to it *only after all the other possibilities to explain the value of the cosmological constant by a fundamental theory have been exhausted.* At present, this is far from being the case. For example, it is possible that the cosmological constant, or the energy density of the vacuum, is in fact not constant after all, but rather varies with time (possibly even in relation to some other physical quantity). An idea along these lines was explored by Princeton cosmologists James Peebles and Bharat Ratra a decade ago and, in a more recent incarnation, based on string theory concepts, by Robert Caldwell and Paul Steinhardt of the University of Pennsylvania. The important point is that because of the universe's old age, it is easy to envisage how a quantity that varies with time could become either very small or very large. For example, if something starts from the value zero and increases by a mere one ten-billionth every year, it will reach the value of one in 10 billion years.

Interestingly, more than a decade ago, cosmologist Lawrence Krauss of Case Western Reserve University suggested the name "quintessence" for the dark matter in galaxies and clusters of galaxies. He borrowed this term from Aristotle, who used it for the ether—the invisible matter that was supposed to permeate all space. Somehow the name did not stick. Now Caldwell and Steinhardt propose the term "quintessence" to describe the variable energy that may represent a variable cosmological constant. At the time of this writing, it is far from clear whether quintessence or any similar concept will provide an answer to the value of the cosmological constant (assuming the latter is correct). What is clear, however, is that *the efforts to explain it in terms of a fundamental theory continue,* and as long as one cannot prove unequivocally that these efforts are doomed to fail, explanations based on anthropic reasoning may be premature. Furthermore, there exists a slight danger (or suspicion) that in spite of its scientific credentials, the anthropic principle still represents the unconscious vestiges of anthropocentric thinking. An amazing example for such unconscious "centering" was provided in 1998 by a discovery from the arts. Christopher Tyler, a neuroscientist from the Smith-Kettlewell Eye Research Institute in San Francisco, examined many paintings in order to determine whether the different functions of the left and right sides of the brain might somehow be reflected in them. To his astonishment, he found that in portraits, a vertical line drawn down the *center* of the painting often fell right on one of the sitter's eyes. Gathering portraits executed by 265 artists from the past six centuries, he found that two-thirds of the paintings had an eye within 5 percent of the center of the frame's width, and one-third had an eye precisely at the center! His sample included, for example, the *Mona Lisa* and the famous *Portrait of a Lady* by Rogier van der Weyden (in the National Gallery of Art, in Washington, D.C.). Tyler then tried to find a specific instruction from any artist to any student to place the eye in the middle of the painting. His failure to find any such instructions led him to conclude that this eye-centering was carried out unconsciously, in response to some yet unknown aesthetic need operating in the brain. I definitely do not mean to imply that the same need is responsible for anthropic reasoning, just to point out that our brain is apparently conducive to self-centering.

In the Eye of the Beholder

Another possibility, in principle, to solve the apparent present conflict between beauty and the cosmological findings is that our ideas of beauty themselves will change in the future. This could even happen in two distinctly different ways. One is that our notions of what is truly fundamental in the theory will change. We have seen examples for such changes of opinions in the history of science. The orbits of the planets were assumed to be circular because of the confusion between symmetry of shapes and symmetry of the laws. Similarly, the fact that there appear to exist several types of dark matter, baryonic and nonbaryonic, may be taken to be either ugly or unimportant for the beauty of the ultimate theory, depending on the fundamentality one associates with dark matter components. The cosmological constant (or density of the vacuum) could turn out to be less significant (and potentially calculable) than we presently think in the grand scheme of things. In such a case its particular value would be of no consequence for the beauty of the final theory. Rather, the truly *fundamental* quantity would be *omega (total)*, which has to be equal to 1.0 (if the inflationary model is correct), and how that value is achieved (from ordinary matter, exotic matter, and the vacuum) would be of no importance. Interestingly, the history of the cosmological constant itself also contains a few examples of extremely opposing views as to the importance of the constant (long before its value had been determined). Arthur Stanley Eddington, for example, who was the most distinguished astrophysicist at the beginning of the twentieth century, declared in his book, *The Expanding Universe*, that "I would as soon think of reverting to Newtonian theory as of dropping the cosmical constant . . . To drop the cosmical constant would knock the bottom out of space." By contrast, the famous theoretical physicist Wolfgang Pauli (after whom the Pauli exclusion principle in quantum mechanics is named) expressed the view (in his book *Theory of Relativity*) that the cosmological constant "is superfluous, not justified, and should be rejected."

A second way in which our perception of beauty may evolve is that the conventional wisdom concerning what "beauty in physics" means will change (not to speak of the fact that some physicists may simply disagree with my definition). An interesting study that was published in November 1998 exemplifies the fact that changes in the views on beauty in general can occur. The study, by Douglas Yu from Imperial College at Silwood Park and Glenn Shepard Jr. from the University of California at Berkeley, examined the question of why some humans are considered more beautiful than others. Theory suggests that attraction to beautiful individuals merely represents an adaptation for choosing mates displaying characteristics indicative of fertility. In particular, it has been claimed that male preference for women of *low* waist-to-hip ratio is culturally invariant, and this has been taken as evidence for the adaptationist interpretation. This conclusion is based on the fact that it was found that infertile women and women suffering from other health disorders tend to have a *high* waist-to-hip ratio. The researchers noted, however, that all the cultures that have been previously tested have in fact been exposed to Westernization by the media. In their new study, therefore, Yu and Shepard assessed the waist-to-hip ratio preferences in the isolated population of the Matsigenka indigenous people in southeast Peru. Access to this region (the core of Manu Park) is restricted to all but official and scientific visitors. To their surprise, the researchers found that the preferences of males in this isolated population differed strikingly from those in the U.S. control population, as well as from all previous results. The highest ranked females (in terms of attractiveness, health, and preferred spouse) were those that were overweight, and within weight classes, *high* waist-to-hip ratio women always ranked significantly above *low* waist-to-hip ratio ones. A study among the much more Westernized population of Alto Madre males gave results that did not differ significantly from those obtained in the United States. The researchers concluded that "a fuller evolutionary theory of human beauty must embrace variation, rather than focusing on 'universal' traits to the exclusion of cultural effects." Thus, if views on human beauty can change, perhaps so can views on beauty in physics.

Many years ago I heard a joke about such changes of point of view from a particle physicist at Tel Aviv University. He asked me: "Do you want to hear a brief history of the evolution of Jewish philosophy?" Not knowing what to expect, I replied that of course I would be interested to hear about that. He said: "First there was Moses, and he claimed it was all there," and he pointed with his finger toward the *heavens.* "Then," he continued, "there was King Solomon, the wisest of all mankind, and he said it was all here," and he pointed with his finger toward his *head.* "Then there was Jesus, and he said it was all here," and he pointed toward his *heart.* "Then came Karl Marx, and he said it was all here." He pointed toward his *stomach.* "He was followed by Sigmund Freud, and you know where he said it all was. . . . Finally," he concluded, "there was Einstein, who said that actually it is all *relative* and in the eye of the beholder!"

While the discussion in this and the previous sections does not offer a *specific* solution to the current crisis, it certainly suggests that there is no reason now to rush to abandon the idea of a beautiful fundamental theory.

The "Cosmological Aesthetic Principle"

The recent cosmological findings, if fully confirmed, have, nevertheless, raised serious questions concerning the beauty of the fundamental theory of the universe. While one could adopt a dismissive attitude toward this problem, by claiming that beauty is not a formal requirement that a physical theory needs to satisfy, this was certainly not the attitude of some of the best scientific minds of and past centuries. The French mathematician, physicist, and philosopher of science Jules Henri Poincaré (who lived from 1854 to 1912) went so far as to state that beauty and the quest for beauty is the *main goal* of scientific endeavors. In his words:

> The scientist does not study nature because it is useful to do so. He studies it because he takes pleasure in it; and he takes plea-

> sure in it because it is beautiful. If nature were not beautiful, it would not be worth knowing and life would not be worth living.

Subrahmanyan Chandrasekhar, one of the most outstanding astrophysicists of the twentieth century, and after whom the x-ray telescope "Chandra" (launched in July 1999) was named, gave over a period spanning forty years a series of seven lectures that have been collected into a book entitled *Truth and Beauty*. Somehow, although I knew of its existence, I never came around to reading that book. As I was preparing to write this last chapter, I decided to browse through his book. Being very familiar only with Chandra's (as he has always been known) technical books on topics in astrophysics and hydrodynamics, I was astonished to see some of the contents of this book. Chandra's specialized books are characterized by containing almost exclusively rigorous and complex mathematical manipulations. Here was a book that contains numerous quotes from distinguished scientists and writers (including the one above by Poincaré) on the importance of the quest for beauty in science. I found one of these quotes, by J. W. N. Sullivan, who wrote biographies of Newton and Beethoven, particularly striking. It reads:

> Since the primary object of the scientific theory is to express the harmonies which are found in nature, we see at once that these theories must have an aesthetic value. The measure of the success of a scientific theory is, in fact, a measure of its aesthetic value, since it is a measure of the extent to which it has introduced harmony in what was before chaos.
>
> *It is in its aesthetic value that the justification of the scientific theory is to be found* [emphasis added], and with it the justification of the scientific method. . . . The measure in which science falls short of art is the measure in which it is incomplete as science! [Exclamation point added.]

Here was a better expression than I could ever have precisely formulated of my thoughts. In spite of the apparent current problems,

I would like to believe that the beauty of the fundamental theory will eventually prevail. In his book *Dreams of a Final Theory,* the great particle physicist and cosmologist Steven Weinberg concludes: "We believe that, if we ask why the world is the way it is and then ask why that answer is the way it is, at the end of this chain of explanations we shall find a few simple principles of compelling beauty."

In physics, certain statements of a more or less axiomatic status are dubbed "principles." The best-known and most fundamental example of these is the principle of least action. The "action" represents typically the difference between two physical quantities; for example, in relation to Newton's laws of mechanics, the action is the difference between the kinetic energy and the potential energy. The principle of least action states that the action stays as small as possible under all circumstances. In the above example, minimizing the action leads to what is known as Newton's second law of motion, which states that the force applied is equal to the object's mass times its acceleration.

In the present book we have encountered three "principles" in relation to cosmology: the cosmological principle, the Copernican principle, and the anthropic principle. We saw that in spite of the fact that the cosmological principle (that the universe is homogeneous and isotropic) had originally been postulated by Einstein somewhat axiomatically (and formulated as a principle in 1935 by British cosmologist Edward Arthur Milne), modern observations confirmed its validity in our observable universe. The Copernican principle in the form of "we are not center stage" originally was applied only to our physical *location* in the universe being typical. In this book, I generalized the principle to include the *time of our existence,* the *matter* we are made of, *the status of our entire pocket universe,* and, according to the most recent findings, *the significance of matter in general* (compared to the vacuum). The recent discoveries in cosmology thus definitely take us through a "second Copernican revolution."

The anthropic principle is still not quite deserving of a "principle status" in my opinion, since although it may certainly turn out to be a powerful tool in delineating the scope and ultimate limitations of the fundamental theory, this still needs to be demonstrated. Indeed, the great Cambridge cosmologist Martin Rees advocates the use of the term "anthropic reasoning" rather than "principle."

I now come to what in some sense is the most difficult part of this book. I have debated long and hard with myself whether I want to suggest a new "principle"—the *cosmological aesthetic principle.* The danger in such a suggestion is, of course, that the name may sound unduly pretentious. A close examination of the history of physics and cosmology reveals, however, as I think this book abundantly shows, that physicists have *in fact long adopted this "principle" wholeheartedly,* and it is only its name that, as far as I can tell, has not been formally spelled out. I therefore propose that fundamental theories of the universe should satisfy the cosmological aesthetic principle, meaning simply that they should be *beautiful.* I have made in this book a serious attempt to characterize what I mean by "beautiful"—that the central idea should be based on symmetry, simplicity (reductionism), and the generalized Copernican principle. While it is quite possible that other physicists will modify somewhat my definition of beauty, I hope that few will disagree with the fact that physics assumes and has essentially always assumed an underlying cosmological aesthetic principle.

John Archibald Wheeler, whose pioneering research on the nature of space-time and on the foundations of quantum mechanics inspired generations of physicists, said once that "observers are necessary to bring the universe into being." By this he meant that only a universe that contains conscious observers has a real existence. This is known as the participatory universe. Most physicists do not accept this point of view, especially since one would like an all-encompassing theory of the universe to include the observers as an essentially inevitable outcome, rather than as disjoint entities.

There exists, however, a more philosophical, rather than physical, sense in which Wheeler's views give a true description of our universe and of our role as conscious observers within it. Assume for a moment that eternal inflation is correct and that our pocket universe is but one of a huge ensemble of universes. Imagine now that we are zooming in, in both time and space (and indeed even from one time to another), first by traversing a whole series of pocket universes until we reach our own pocket, and then inside our own universe through its time history and across structures of different sizes, and finally from the dark matter to a supercluster, to a cluster, to the Milky Way, to the solar system, to Earth, to us, to the neurons in our brain, to molecules, atoms, protons, and quarks. Given the gigantic zoom power required for such a journey, one might conclude that humans are nothing but an insignificant speck of carbon dust in this immense splendor of nature. However, we could conduct our travel plans differently, by merely sitting over the centuries inside the human brain and traveling to wherever new discoveries lead us. Not surprisingly, we would have found out that this journey would have taken us to precisely the same spots as the zoom-in journey. The point is simple: we know that the earth is not the center of the universe because *Copernicus discovered that.* We know that the solar system is not at the center of our galaxy because *Shapley discovered that.* We know that our universe is expanding because *Hubble discovered that.* Finally, we know that protons are composed of quarks because *Gell-Mann discovered that.* At every step, the expansion of our universe was at the same time an expansion of the human horizons of knowledge. In this sense you could almost say that (from our point of view) the universe was not expanding until we discovered that it is expanding. Thus, even though our *physical* existence may look less and less central with the increased applicability of the generalized Copernican principle, from our vantage point as observers we *are* center stage in *our* universe!

Maria Mitchell, the pioneering American astronomer of the

nineteenth century (whose life is beautifully described in the book *Sweeper in the Sky,* by Helen Wright) wrote the following words, which I find most fitting to end this book.

> These immense spaces of creation cannot be spanned by our finite powers; these great cycles of time cannot be lived even by the life of a race. And yet, small as is our whole system compared with the infinitude of creation, brief as is our life compared with cycles of time, we are so tethered to all by the beautiful dependencies of law, that not only the sparrow's fall is felt to the outermost bound, but the vibrations set in motion by the words that we utter reach through all space and the tremor is felt through all time.

Index

Index